**WOOD HEAT
SAFETY**

WOOD HEAT SAFETY

by Jay W. Shelton

GARDEN ✿ WAY PUBLISHING
Charlotte, Vermont 05445

Illustrations by Stephen M. Lynch
All photos by the author unless otherwise indicated.

Printed in the United States
Fourth printing, August 1981

Library of Congress Cataloging in Publication Data

Shelton, Jay, 1942–
  Wood heat safety.

  Bibliography: p.
  Includes index.
  1. Stoves, Wood—Safety measures. 2. Wood as fuel
—Safety measures. I. Title.
TH7438.S5   697'.04   79-17951
ISBN 0-88266-160-4

Dedicated to Katherine, Jeremy, and Reed,
and all others who enjoy, or at least tolerate,
heating with wood.

Contents

Preface

For a century interest waned in the use of wood as the source of heat for the home. The convenience of heating with coal, and then with oil and gas was persuasive.

Then, in the early 1970s, the price of oil rose, and with it the price of coal, gas, and electricity. And as if the price increases were not enough, actual and threatened interruptions of the supply were increasing. Thus quite suddenly wood became an alternate fuel to consider.

Can wood heat a house?

Some were skeptical, but the answer was apparent to many. Old wood stoves, rusty after years of disuse, were polished up and carried back into homes. Homeowners dug through years of wallpaper to find entrances into flues. Other stoves were hooked up through fireplaces, or into new insulated metal chimneys. Wood heat was back, and not miraculously, it still worked.

But can wood do the job efficiently and economically?

It all depends.... For many the answer is clearly "yes!" For those who live in or near wooded, low population density areas, and who do not mind doing some of the work to prepare their own fuel, wood heating clearly can be economical. Those who supplement electric heating with wood almost always save money, even if they purchase seasoned, split, and delivered wood. Of course the wood must be burned in an energy-efficient unit—and this does not include the traditional open masonry fireplace.

The comfort of wood heating can be great. Many people are discovering the pleasure of a "hot spot" in the home. And, in an insulated and tight house, one can obtain remarkably uniform heat throughout from a single stove. Increasingly popular are wood- and coal-fired *central* heaters— furnaces and boilers—whose heat is distributed via ducts or pipes to heat an entire house to as uniform a temperature as any conventional heating system could do it.

Now a third question is heard.

Is heating with wood (and with coal) safe?

The question is repeated in nagging fashion as our newspapers each winter report homes burned, and "wood stove" is mentioned in the account.

As I try to answer the question, several points become clear.

1. Many of those using solid-fuel heaters have not been trained in their safe use because they were brought up in homes with electric, gas, or oil heat.

2. Many building codes do not cover solid-fuel heating adequately and reasonably because there was no reason to, so few were the installations until recent years.

The answer to the safety question is, "Probably—if."

This book explains that "if." Above all else, it provides clear information on how to use wood and other solid fuels *safely* for heating. The bulk of the recommendations are derived from building codes and from the National Fire Protection Association (NFPA). The many illustrations should help make this information easy to understand.

But this book goes beyond interpreting this standard safety information. It discusses the "whys" to give a deeper understanding of the safety issues and hence an awareness of what is critical and what is peripheral.

Where in my opinion building codes are clearly inadequate or excessive, this book discusses new or alternative standards. It also provides extensive material on *safe operation and maintenance of systems*, an area not covered fully or accurately in many sources of information.

This book should help consumers select, install, operate and maintain solid-fuel heating systems safely. We hope it will also assist building and fire prevention officials to be constructively and flexibly helpful to homeowners by helping these officials to gain a better understanding of the *intent* and *rationale* of building codes, and awakening an awareness of safety issues not addressed in building codes.

This book contains extensive material on the recommendations of the National Fire Protection Association (NFPA) and the requirements of many building codes. This material is not "official" in the sense of being direct quotations, or of having been reviewed and verified by NFPA or building code authorities. Although the author believes these representations to be correct, in cases where accurate renditions are critical, the primary sources should, of course, be consulted.

In localities with building codes, the only *legal* installations are those in compliance with those local building codes.

Acknowledgments

I am indebted to Larry Gay, John Rummler, Paul Stegmeir, Richard Stone, Peter Swift, David Park, and Allan Dudden for their help and suggestions concerning the content of this book; to Stephen M. Lynch for his patience in working on the illustrations; to TGL Photoworks for excellent technical assistance; and to the Massachusetts State Fire Marshal's Office for assistance in supplying fire statistics.

CHAPTER 1

The Problems

Imagine a typical winter's scene in a country home. The family members are sitting around the fireplace after dinner. There's been talk, but now it's getting late and they're staring into the depth of the flames, each alone with his thoughts.

This is a scene of serenity, and certainly one that reflects to most of us the comfort and safety of the home.

Comfort, perhaps, but not necessarily safety.

If there is a fire, be it in a fireplace or a stove, there's danger.

HOW FIRES START

Wood heating equipment can start uncontrolled fires in two fundamental ways.

One is by the wood fire itself getting outside the stove, fireplace, or chimney and igniting the house (Figure 1–1). There are many examples: flames leaping out of a fireplace, burning logs rolling from a Franklin stove, sparks jumping out an unbaffled air inlet damper, ashes, still glowing, setting fire to a cardboard box, flames or burning creosote pushing through cracks or faulty joints in a chimney, and sparks wafted from a chimney and landing on a roof.

The other way is when combustible materials are too close to the wood heating system (Figure 1–2). This can result in spontaneous combustion due to high temperatures alone. Examples are placing a stove too close to a combustible wall, placing furniture too close to a stove, drying wood

on a stove top, providing inadequate protection when installing a stovepipe through a wall, leaving too little space between chimneys and combustible parts of the building, and permitting dangerously hot air from a supplemental wood furnace to enter an ordinary duct system.

There is evidence that temperatures as low as 200°–250° F. can cause wood to ignite (Appendix 1). Years of exposure increase the possibility, and the evidence indicates that high or varying humidity and varying temperature may contribute to the danger.

While the probability of a fire at such low temperatures has not been established and may be rather low, safety standards, being appropriately conservative, aim at keeping combustible materials below about 200° F. most of the time.

Fire is not the only safety problem linked with wood heating. Asphyxiation results from smoke originating from a malfunctioning wood or coal heater or a chimney, since that smoke contains toxic substances such as carbon monoxide. Smoke can also cause death because of its temperature, its low oxygen content, or the solid and liquid particles in it.

Skin burns occur when the hot parts of a wood heater are touched. While such burns can be painful, they are rarely a serious medical problem.

Steam explosions are a hazard in water heating systems that are improperly designed. Their frequency is rare, but their consequences can be serious.

Finally, there are the many hazards faced when getting the wood supply. Use of chain saws, axes,

1

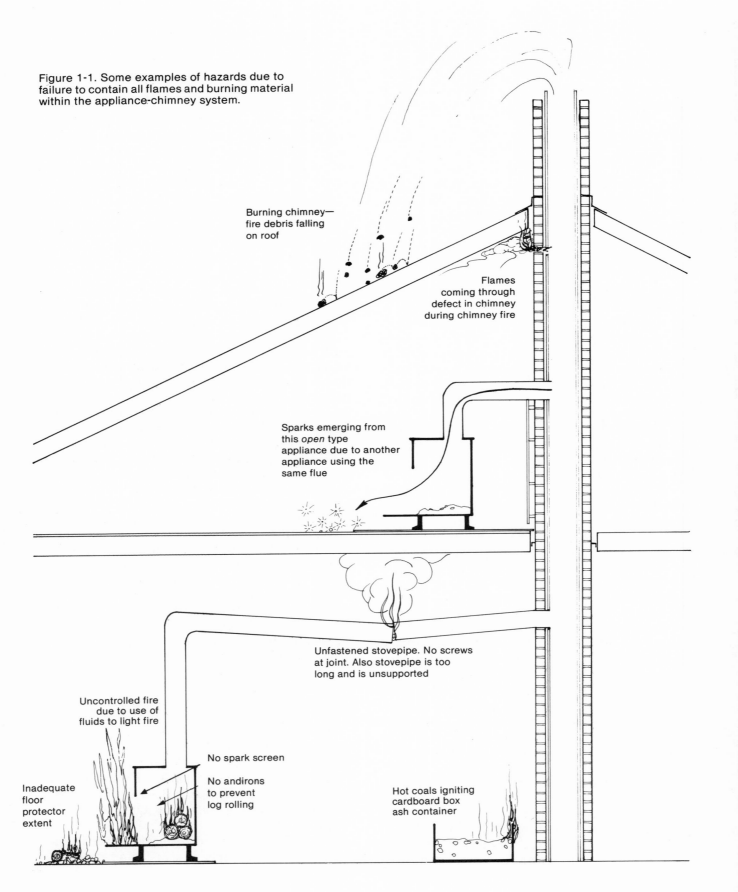

Figure 1-1. Some examples of hazards due to failure to contain all flames and burning material within the appliance-chimney system.

Burning chimney—fire debris falling on roof

Flames coming through defect in chimney during chimney fire

Sparks emerging from this *open* type appliance due to another appliance using the same flue

Unfastened stovepipe. No screws at joint. Also stovepipe is too long and is unsupported

Uncontrolled fire due to use of fluids to light fire

No spark screen

No andirons to prevent log rolling

Inadequate floor protector extent

Hot coals igniting cardboard box ash container

splitting implements, and heavy power equipment in forests involves risk. The high cost of workmen's compensation insurance in the forestry industries is evidence of the hazardous nature of these activities even when done by professionals.

Is Heating With Wood or Coal Dangerous?

In all honesty, the answer is not known at this time. Statistics are needed but not available on the difference in risk of fire for homes with and without wood or coal heating.

Nationwide, insurance company payments for fire damage related to wood heating are very small compared to other causes of fires. But the percentage of homes heated with wood is so small that this proves nothing.

Insurance firms claim that in some local areas, the number of house fires has doubled due to the use of wood stoves. If proven true, this *is* alarming.

The *causes* of fires are somewhat better understood. The information is encouraging since it indicates the vast majority of wood-stove-related fires are avoidable. Two studies agree in their conclusions: *heater-related fires are caused almost exclusively by installation, operation, and maintenance errors, not by unsafe equipment.*

During the 1977–78 heating season, 104 wood-heating-related fires were reported to the Massachusetts State Fire Marshal's Office.[1] Local fire departments blamed 74 of these on a cause more specific than "wood stove" or "overfired wood stove." Of these 74, a total of 71 percent were attributed to faulty installation, 16 percent were apparently due to poor maintenance, such as chimneys not kept clean or furniture moved too close to the stove, 11 percent were attributed to operator errors, and one single fire was blamed on a "defective stove," with no details given. Since the house burned to the ground, it is not clear how the determination was made that a defect in the stove caused the fire (Table 1–1).

While some of the local fire department investigations were incomplete, the message was clear: *if a wood heater is installed properly and is carefully operated, and the installation is conscientiously maintained, the fire hazard is reduced to a very low level.* A second study,

summarized in Table 1–2, supports this conclusion, as does a recent study by the National Bureau of Standards (see Bibliography).

I expect that heating with wood may always be a little more dangerous than heating with systems using gas, oil, or electricity. One reason is that oil and gas heaters have built-in automatic safety features. Second, machines are often more reliable than people. Wood heaters require much *human* attention, and humans always have and always will make mistakes. Finally, chimney fires are dangerous, and occur more often with wood and coal heating systems, despite conscientious cleaning programs.

Having read all of the above, should your decision be to avoid wood heating? If you expect wood heating to be as simple and safe as an occasional adjustment of a thermostat, then stick to electricity, gas, or oil. For solid-fuel heaters, equivalent safety requires somewhat more effort, and respect for fire. My personal response is to enthusiastically heat my own home with wood, not because there is no risk, but because when done carefully, heating with wood probably involves no more risk than do many other accepted choices and activities in life, such as living in a region with polluted air, and travel. What is more, heating with wood can save money, can keep you warm despite embargoes, strikes, and blackouts, and it is fun.

Why Be Safe?

Despite such potentially serious consequences as loss of life, injuries, and loss of property, many individuals shortcut safety in the installation and operation of wood heaters, in an effort to save money or time. It is very human not to think of yourself as a statistic—to think you will be safer than the "average" person—that it is only other people's houses that burn down. Whatever an individual's feelings are about shortcutting safety and the consequent higher risk of fire, some indirect consequences should be understood.

First, if houses are close together or you live in an apartment building, even if you want to risk your own life and property, you should not because you are exposing others to risk. Society has a very clear right to force you to be safe.

Second, if you and a few others are shortcutting safety, it will still cost you and other people money even if you personally win the gamble and do not have a fire. Statistically, installing and operating wood heating equipment unsafely results in more fires, even if it is not your house that burns. Local

---

1. The interpretation and analysis of these data are by the author. This analysis is also published as J.W. Shelton, "Analysis of Fire Reports on File in the Massachusetts State Fire Marshal's Office Relating to Wood and Coal Heating Equipment," NBS-GCR-78-149, National Bureau of Standards (1978).

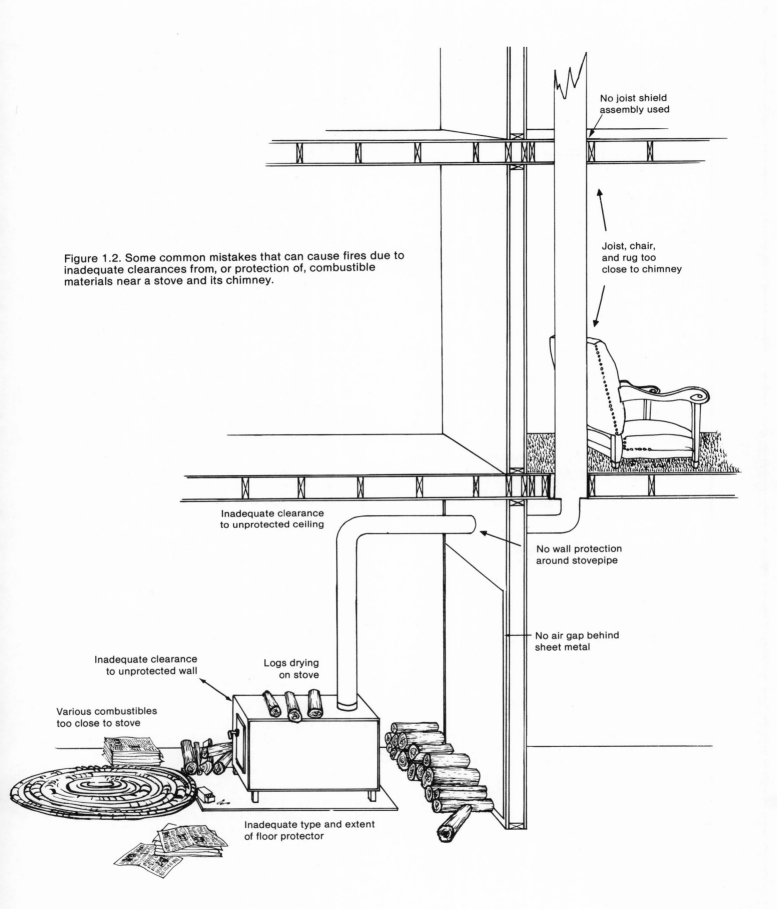

Figure 1.2. Some common mistakes that can cause fires due to inadequate clearances from, or protection of, combustible materials near a stove and its chimney.

No joist shield assembly used

Joist, chair, and rug too close to chimney

Inadequate clearance to unprotected ceiling

No wall protection around stovepipe

No air gap behind sheet metal

Inadequate clearance to unprotected wall

Logs drying on stove

Various combustibles too close to stove

Inadequate type and extent of floor protector

fire departments will have to fight more fires, and this costs money. Health, property, and life insurance premiums will rise as long as people continue to heat with wood unsafely. Thus we *all* pay these costs of the extra risks taken by a few.

The home owners' insurance situation may develop into an even more compelling reason to use wood heaters safely. Some insurers now will not issue a policy unless the installation has been approved by the local building inspector or fire department.

Other insurance companies are considering a higher premium for houses with wood heating. But they may also offer discounts for a verified safe installation, a smoke detector system, and annual chimney cleaning and general inspection.

There is even a chance that with your current policy, you may not be able to collect any money if your house is damaged by a wood-heater-related fire if the installation is illegal because you did not obtain a required building permit, or you did not notify the insurer that you were heating partially or fully with wood. Most insurance policies require that you inform the insurer of certain types of changes in the house, such as in the heating system.

To be on the safe side, you should notify your insurance firm in writing when you install a wood heating system.

In summary, installing and operating a wood heating system safely (and this includes abiding by local or state regulations) not only may save your life and property, but it will certainly reduce the costs to all of us of fire fighting and various insurance coverages. And safe installation may be necessary in order to have a valid home owners' insurance policy.

Safety—A Matter of Degree

There is no such thing as an absolutely safe installation, nor an absolutely safe operator, nor an absolutely safe wood heating appliance.

As an example, consider floor protection in front of a fireplace stove. Most codes require that protection over combustible floors extend at least 18 inches beyond the door side of stoves. Presumably most flying sparks or fallen coals will not travel further than 18 inches. Of course some will. But then should not the protector extend further than 18 inches, say, to 24 inches? That would certainly be safer—there might then be only three fires a year due to flying sparks instead of thirty. With a 36-inch protector, there might be only one fire every ten years from this cause. (These

Figure 1-3. Scars on a floor from hot coals and sparks, illustrating the arbitrariness of any particular floor protector size. Most building codes require an 18-inch extent in front of a stove. (Photo by the author, with the cooperation of V. G. Collins.)

statistics were made up to illustrate the probabilistic nature of safety.) No reasonable floor protection system is perfect. Some day, somewhere, a very energetic big spark will set a new record for distance traveled from the stove and start a fire.

Of course careful operation of a stove can decrease the chances of sparks getting out of a fire. Screens should be used over fireplace stoves whenever the doors are open, and extra vigilance should be used when burning wood that tends to spark and pop. But here again safety cannot be absolute. The screen cannot be on *all* the time, particularly during fueling or marshmallow toasting. If one is enjoying an open fire without the screen, a sudden emergency elsewhere in the house may result in the fire being left unscreened.

Table 1-1. Massachusetts Safety Study

REPORTED CAUSE[1]	No. of Cases	Approximate Percentage of All Reported Fires	Approximate Percentage of Fires for Which a Usefully Specific Cause was Given
Improper Installations			
Not specific	6		
Stove			
not specific	7		
clearances	8		
Stovepipe connector			
not specific	7		
clearances	9	53 cases	51 71
joints not fastened	2		
Chimney			
not specific	7		
clearances	3		
cracked chimney	1		
no (or inadequate) liner	3		
Lack of Maintenance			
Creosote buildup and/or chimney fire	11	12 cases	11 16
Clearances to furniture not maintained	1		
Negligent Operation			
"Wood shingles drying on top of wodd stove."	1		
"Door may have been left ajar."	1		
"…new matress…was leaning against the kitchen wall next to a wood burning stove. Evidently a spark from the stove ignited the matress, which in turn ignited the rug and floor."[2]	1		
"Possibly started in area where ashes from a wood stove were dumped. Wood ashes stored in a cardboard container."	1	7 cases	7 11
"Spark from wood burning stove ignited overstuffed chair and scorched parlor floor."[2]	1		
"Overheated wood stove and overheated flue pipe. Woman of house filled stove with wood and went to town, leaving damper wide open…Stove was very good model and the installation appeared to be satisfactor. I must believe that this fire was caused by human error other than faulty installation or equipment."	1		
Defective or Unsafe Equipment			
"Defective wood stove. Home completely destroyed."	1	2 cases	2 2
"Steam explosion of water heating jacket in antique cook stove."[3]	1		
Other (e.g. "Wood stove.")	30 cases	29	
	104 cases	100	100

1. In cases where two or more possible causes were given or implied, I have selected what I consider the more fundamental cause. For example, one report reads "Overheated chimney connector from wood stove ignited combustible wall." I interpreted this as a case of inadequate clearance between the stovepipe and the wall, since with proper clearances, even overheated stovepipe is very unlikely to ignite comubustible walls.

2. I have assumed sparks got out because a spark screen was not in use when the doors of a Franklin stove were open. Other interpretations are: (1) inadequate clearances to walls and furnishings, and (2) unsafe stoves that let sparks out with doors shut.

3. This case might also reasonably be interpreted as due to operator error—firing up a stove with a capped water jacket with water in it.

Fires and other accidents related to the use of wood heating equipment in Massachusetts on file in the State Fire Marshal's Office as of June 26, 1978, covering most of (and mostly) the 1977–78 heating season. The total sample is 104 reports, 74 of which gave usefully specific causes. Determining causes of fires is difficult, particularly when the building is destroyed. Many of these local fire department reports are only educated guesses.

Table 1-2. Wisconsin Safety Study

Percentage	Causes
35%	Unsafe chimneys (10 percent were exterior concrete ring-block chimneys).
20%	Inadequate clearances of stove or stovepipe from walls, ceilings and floors.
10%	Ash disposal into combustible container.
5%	Sharing a flue with a gas or oil appliance.
5%	Use of flammaable liquids for starting fires.
5%	Melting of stainless steel chimney liners.
20%	Miscellaneous, including chimney fires with no resulting house fire.

Summary of an insurance company study of wood heating related fires in portions of the Midwest. (Private communication from American Family Mutual Insurance Co., Madison, Wisconsin.)

And of course guests and babysitters may not be as careful as the usual operator. Thus house fires can start from flying sparks despite the best of intentions of the owners and operator.

Adequate maintenance is also required to minimize the chance of sparks causing fires. All combustible materials must be *kept* away from the stove and off the floor protector, including rugs and carpets, pillows, mattresses, newspapers, and curtains.

The wood heating appliance itself also cannot be *perfectly* safe against this or any other hazard. Baffled air inlets on stoves reduce substantially the chances of sparks getting out during closed-door operation, but do not cut those chances to zero. Spark screens will not stop very small sparks from getting through the screen mesh. Spark screens rarely fit spark-tight at their edges against the fireplace opening. Many spark screens can be knocked over by shifting wood in the fire chamber.

Safety weaknesses in any of the four areas of installation, operation, maintenance, and equipment design can be compensated for by extra precautions in one of the other areas. It is relatively safe to leave off spark screens if the floor protection is unusually large in extent. Having no floor protection can be relatively safe if the stove has a spark baffle in the air inlet, is used only with its door or doors shut, and is used only in the presence of a very watchful operator. A smoke detector in the room with the stove or fireplace can help prevent loss of lives and serious damage to the building in cases where preventive measures have been inadequate.

There are other examples of the relativity of safety. Most building codes require radiant stoves to be installed at least 36 inches away from unprotected walls containing wood. Will a stove at 30 inches cause a fire? No, not usually. The 36 inches is selected as being safe for the biggest and hottest stoves. Smaller stoves *are* safe at less clearance than large stoves. Even large stoves can be safe at less than 36 inches if design features or operating habits keep their surface temperatures down somewhat.

Pressure relief valves are required (and needed) on most systems involving heating water, to prevent steam explosions. The valves prevent the vast majority of potential explosions, but not all. The valves can fail. Valve checking and maintenance can help.

Clearly safety is a matter of degree, not a black-and-white issue. To achieve the highest degree of safety one should take reasonable precautions in selection, installation, operation, and maintenance of equipment. And since so many wood-heating-related fires are started by negligence by the operator, *complacency* that the appliance itself and the installation are safe is the wood heater's most dangerous enemy.

HOW TO HEAT SAFELY WITH WOOD

Heating with wood or coal safely requires safe equipment, a safe installation, and safe operation and maintenance.

Equipment should be either:
"Approved," meaning "acceptable to the authority having jurisdiction," for example, acceptable to the local building inspector; or
"Listed," meaning "included in a list published by a recognized testing laboratory or inspection agency, indicating that the equipment meets nationally recognized safety standards." A well-known laboratory that tests and lists many kinds of equipment is Underwriters Laboratories (UL).

If neither of the above is available, one can read the portions of this book on equipment safety and make an independent judgment. Most equipment available today is reasonably safe.

Increasingly, building codes are *requiring* that wood- and coal-burning heating systems be "approved" or "listed."

7

Installations will be relatively safe if one of the following is observed:

1. For listed equipment, installations should follow the manufacturer's instructions. The process of listing includes careful attention to instructions.

2. For unlisted equipment you can follow the supplied instructions, but it is wise to consult the recommendations in one of items 3, 4, and 5 below. If the recommended installation guidelines in one of these other sources differ from those provided by the manufacturer, and they are more conservative, deviations from the manufacturer's instructions may be appropriate.

3. Follow local building code requirements.

4. Follow the guidelines of the National Fire Protection Association (NFPA). Most fire safety codes are taken from NFPA recommendations. NFPA's recommendations are contained in NFPA publication numbers HS-10, 89M, 90B, and 211. See Bibliography.

5. Follow the guidelines in this book. The principal features of most building codes and of NFPA's recommendations are explained and illustrated extensively in this book, and so are my own independent judgments.

The guidelines for safe installations in items 2 through 5 are not generally equivalent. Instructions with unlisted equipment can be deficient, and this book differs from NFPA doctrine by covering more aspects of installations and by being more conservative in some areas and more liberal in others. These differences are partly a consequence of the inherently statistical nature of safety. The differences are usually not critical. Careful adherence to any of the above sources of installation guidelines with the occasional exception of item 2 will result in a reasonably safe installation.

Operation and maintenance will be relatively safe if one reads, understands and follows the instructions in this book, particularly Chapter 3 and parts of Chapter 5.

In most places, installations will be *legal* only if they comply with the local building code. When equipment is listed, most codes require that it be installed according to the manufacturer's instructions. The detailed provisions in most codes on such subjects as clearances are intended for unlisted equipment and are modeled closely, but not exactly, on NFPA's recommendations. Many building codes require a building permit before installing any kind of wood-burning equipment in a building.

CHAPTER 2

Installations

Most building fires caused by use of wood heating equipment would not occur if the installation met nationally recognized minimum safety standards. The most common installation errors involve the venting system—the chimney and the chimney connector or stovepipe (Figure 2–1).

CHIMNEYS

A chimney should vent safely the smoke and fumes from combustion out of the building, and supply the needed draft to the appliance. Safe venting involves getting the smoke outside without causing a house fire. Chimneys should be well enough constructed to do this through years of normal use.

Chimneys can serve as significant additional sources of heat, and masonry chimneys can also store heat. Since wood and coal heaters tend to generate considerable smoke, the precursor of creosote, a chimney should help minimize creosote accumulation.

Capacity

The capacity of a chimney is the maximum flow of smoke or stack gases that it can safely handle. If chimney capacity is too small for an attached wood heater, either the heater will not be able to get very hot because not enough air is being pulled into it, or smoke will spill out of it, particularly if it is a fireplace or fireplace stove with its doors open. If chimney capacity is too large, the excessive

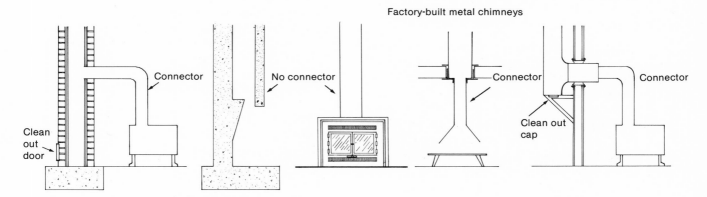

Figure 2–1. Examples of the distinction between chimney connectors and chimneys. Chimneys are the principal portion of the venting system. They are mostly if not totally vertical, are permanently installed, and are designed for decades of normal use. Chimney connectors connect appliances to their chimneys.

Table 2-1. Chimney Sizes

Appliance Type[1]	Collar Size (Inches)	Suggested Inside Diameter for Round Chimneys or Flue Liners (Inches)	Suggested Rectangular Flue Dimensions (Nominal Exterior) (Inches)
Small stoves	4	4–6	4x8
	5	5–7	8x8
Medium and large stoves	6	6–8	8x8
	7	7–8	8x8
Fireplace stoves,[2] small fireplaces[2]	8	8–10	8x12
furnaces, boilers	9	10	8x12
Medium fireplaces[2]	10	10–12	12x12
Large fireplaces[2]	12	12	12x16
	14	14	16x16

1. Each type of appliance spans a range of collar sizes; thus location of type in table is approximate.
2. A common rule of thumb for fireplace and fireplace-stove chimneys is that the cross-sectional area of the flue should be about 1/10 (1/8 for chimneys less than 15 feet tall) of the area of the fireplace opening.

Suggested chimney sizes for residential wood heating equipment lacking manufacturer's instructions. Bigger is not always better. Oversize flues tend to create less draft and accumulate more creosote.

cooling of the smoke can lead to decreased draft and increased creosote accumulation.

Chimney capacity depends most critically on diameter or cross-sectional area. Chimney height and the average temperature of the gases in the chimney are also important. Chimney connectors, usually stovepipe, may reduce capacity of the whole venting system if they are unusually long or have many turns.

Diameter

How do you know what size chimney is appropriate for a given wood heater? If it is a new unit, the manufacturer or importer should provide that information in the instruction manual. If you lack such instructions, the size of the flue collar on the wood-burning unit can serve as a guide. The chimney flue interior diameter should be at least as large as that of the collar, and preferably not more than about 25 percent larger (Table 2–1).

Height

Although increased height in chimneys increases capacity, in practice height usually is not chosen but is determined by such architectural considerations as placement of the wood heater, the height of the roof, and the location and height of the roof peak. If the usual chimney height requirements are followed, venting systems are likely to be at least 10 feet high, measured from the flue collar, and this is usually adequate for stoves. Most factory-built metal fireplaces come with specific instructions on minimum and maximum chimney heights, and on diameters.

Chimneys more than 30 feet high have additional capacity, compared to shorter chimneys, that is equivalent to about one extra inch of flue diameter.

To help avoid possible down drafts in chimneys in windy weather, and to prevent the hot flue gases from overheating the roof, the chimney should extend well above the roof. The usual code-approved minimum height is either 3 feet above the roof where the chimney penetrates, or 2 feet higher than any part of the roof within 10 horizontal feet of the chimney, whichever is larger (Figures 2–2 and 2–3).

In many old New England buildings, such as in the Sabbathday Lake Shaker community in

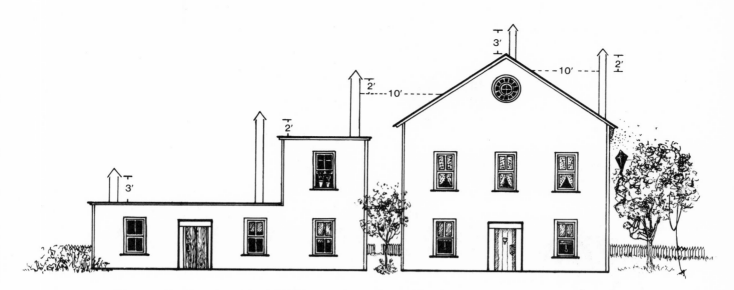

Figure 2-2. Minimum chimney termination elevations according to NFPA and most building codes.

Figure 2–3. Some examples of chimneys that do not terminate high enough above their buildings.

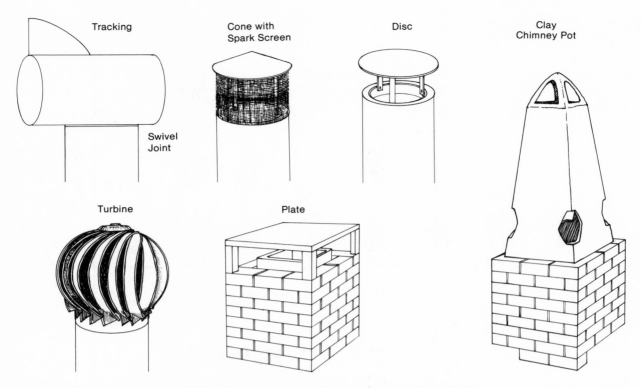

Figure 2-4. Some examples of chimney terminals. Terminating a chimney with a cap can keep rain and snow out and lessen the detrimental effects of wind on draft.

Maine, chimneys are considerably taller, extending at least 6 feet above the roof peak. Such tall chimneys may require extra mechanical support. Also, tall chimneys may be inconvenient to clean and maintain.

No trees should be within about 15 feet of the top of a chimney, particularly at a level with, or above the chimney top. Although trees can affect draft through turbulence they cause in the wind, the more important safety consideration is keeping trees far enough away so that they cannot be ignited by the hot smoke. In dry climates, keeping them more than 15 feet away is essential, especially considering the height and temperature of chimney fire flame columns.

The Canadian building code recommends that chimneys terminate at least 10 feet above the top of any door or window within 50 feet of the chimney, to minimize the possibility of smoke entering a building.

Chimney Caps

Chimney caps (Figure 2–4) are not essential for safety. They keep rain and snow out of the chimney, which can cause corrosion and a mess; improve draft if steady winds tend to angle down into the chimney; and minimize the effects of wind gusts in making small amounts of smoke come out of a fireplace or stove into the house. Caps impede the flow of smoke slightly and thus decrease chimney capacity a little. All things considered, I recommend use of caps.

Caps with moving parts, such as turbine and directional caps, may get stuck—creosote deposits are usually heavy on the cool chimney cap surfaces. Creosote can also quickly plug up caps with less than about 1-inch openings for the smoke to pass through. Such caps should not be used.

I have seen no convincing evidence that specially designed chimney caps can significantly reduce creosote accumulation in chimneys.

Spark and Bird Screens

Spark-arresting screens (Figure 2–4) may be required in some regions where very dry conditions increase the likelihood of forest and brush fires. Spark screens also keep birds and their nests out of chimneys. The screen mesh size must not be so small that it easily clogs with creosote, soot, and ash, and not so large that significant burning or glowing material can get out. A good size is half-inch mesh. To avoid too much flow restriction, especially as the screen gets dirty, the surface area of the screen should be at least four times the flue cross-sectional area. The screen should be checked frequently and cleaned when necessary. A cap and screening may of course be combined.

Chimney Location

The best location for chimneys is *inside* buildings, although experience clearly indicates any location will usually function satisfactorily (Figure 2–5). There are five disadvantages to exterior chimneys. Four of them are a consequence of the colder outdoor environment, resulting in cooler temperatures inside the chimney. Cooler flue gas temperatures result in less draft, more creosote, the possibility of the chimney being non-self-starting, and the possibility of chimney flow reversal. The fifth disadvantage is that the heat given off from the sides of the chimney is wasted, heating the great outdoors.

A non-self-starting chimney has cold outdoor air flowing down it when it is not in use. Flow reversal is essentially the same phenomenon but with a fire in the wood heater—outdoor air flowing continuously down the chimney, feeding the fire, and the smoke coming into the house through the air inlets. Both can occur because heated buildings act like chimneys themselves—they have their own "stack effect."

On a cold day the relatively warmer air in a building tends to rise. The warm air pushes out through cracks in the roof and upper portions of the building, and cold outdoor air is sucked in through any available openings in the lower portions of the building (Figure 2–6). The draft or buoyancy developed in an exterior chimney must at least overcome the stack effect or suction developed by the building in order to operate properly.

The worst case is an exterior chimney serving a wood heater in the basement or ground floor in a multi-story building. Whenever the temperature of the flue gases or air in the chimney is less than the temperature of the air in the house, the direction of flow in the chimney can be downwards.

Exhaust fans in kitchens, bathrooms, and clothes dryers, and operation of any other fuel-burning appliances in a house tend to depressurize the house. This contributes to flow reversal and non-self-starting phenomena. Even *interior*

chimneys *can* be influenced. The negative effect of these other appliances on chimney draft tends to be especially strong in tightly constructed energy-conservative homes with few air leaks. One preventive is to supply direct outside air (see Chapter 5).

When starting a fire this condition is merely an annoyance, and the flow can be reversed by lighting a wad of newspaper that has been inserted up the chimney or stovepipe, or opening an exterior door or window in the room of the wood burner.

Flow reversal in an operating chimney can lead to asphyxiation. It is most likely to occur at night, when outdoor temperatures are lowest, fires are often their slowest, and, unfortunately for them, the occupants are usually asleep.

To correct a reversed-flow chimney in use, one should *open* windows or exterior doors on the ground floor, *close* windows in the upper stories, and *shut off* operating exhaust fans and gas- or oil-fueled appliances. Igniting newspaper inserted in the bottom of the chimney through a cleanout can also be effective. One can also avoid the problem by avoiding relatively cool, smoldering fires.

Use of insulated factory-built chimneys in exterior applications will lessen but not eliminate the probability of flow reversal. In theory, thermosyphon (air-cooled) metal chimneys can be non-self-starting and undergo flow reversal even when installed inside. However the physics of such chimneys is sufficiently complicated that good

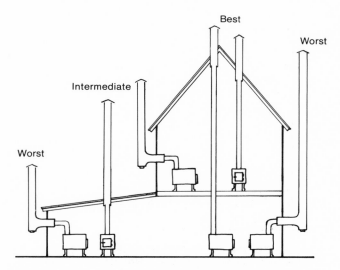

Figure 2–5. Chimney location and draft. Various chimney locations are marked in the figure with respect to draft quality and freedom from flow reversal and non-self-starting problems.

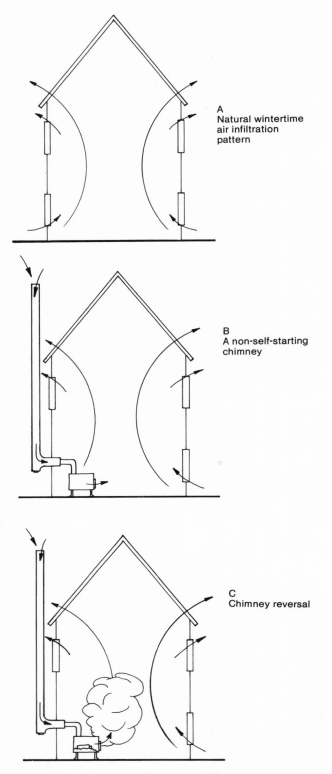

A
Natural wintertime
air infiltration
pattern

B
A non-self-starting
chimney

C
Chimney reversal

Figure 2-6. The stack effect of buildings, and its effect on exterior chimney operation. Normal, calm weather, wintertime air leakage and flow patterns are shown in sketch A. The exposure and hence cooling of exterior chimneys makes them susceptible to flow reversal and non-self-starting.

quality empirical evidence is needed to find out with reasonable certainty.

Especially if you use an exterior chimney, install a smoke detector in the room with the wood heater and near the bedrooms. Smoke detectors are cheap insurance against loss of life.

Ice Plugs

In very cold weather ice plugs may form in exterior chimneys. The problem seems to be most common in small masonry chimneys. The condensed creosote and water drip down the flue liner. They freeze when they get to the cold chimney surfaces below the breaching (where the stovepipe enters the chimney). As additional liquid drips down and freezes, a ring of dirty ice grows in towards the center of the chimney, eventually sealing it off. The ice and liquid build up to the breaching where the liquid may then pour into the connector and come into the house, and the ice can obstruct the flow of smoke from the wood heater.

I know of no simple solution to this problem. Of course, it cannot happen in an interior chimney. Some type of waterproof electric heating could be used to maintain an opening in the ice to let the liquids drain down. But the chimney below the ice plug may then fill up with ice. The liquid needs to be guided to outside the chimney.

Where does the water come from? Each pound of seasoned wood when burned generates more than half a pound of water vapor. A large fraction of this vapor can condense inside a cold chimney. Accumulations of a few gallons each day are possible.

Chimney Cleanout Openings

Chimney cleanout openings at the bottom of chimneys are for easy removal of creosote, soot, ash, and other debris. Cleanouts in factory-built chimneys are typically a capped leg of a T fitting. In masonry chimneys a metal door near the base of the chimney is used. The opening should seal as airtight as possible; leaks will rob the wood burner of draft, and make chimney fires more intense and harder to suppress. Although not directly related to safety, cleanouts have the very important indirect effect of making chimney cleaning more likely by making it easier. A disadvantage of a chimney lacking a cleanout opening is that there is often no place at the chimney bottom for debris to accumulate without blocking the flow of smoke.

There is room for improvement in the cleanout

Figure 2-7. "Ball lightning in a nineteenth century woodcut. The original title, translated from French, reads, 'Ball lightning crossing a kitchen and a barn.' Perhaps the ball lightning came down the chimney used to exhaust the cooking fires. How the young lady's blouse came to be in such a state of disarray is not known." Ball lightning is a fascinating and not fully understood phenomenon. The glowing balls (of what no one is sure) typically last only for a few seconds. They sometimes cause damage. Some sightings associate them with chimneys. The principal reason for chimney lightning protection, however, is to conduct to ground *normal* lightning strokes. (Illustration courtesy of Burndy Library; quotation reprinted from Martin Uman, *Understanding Lightning* [Carnegie, Pa.: Bek Technical Publications, Inc., 1971], p. 124, with permission of the author.)

access in some factory-built chimneys. A press-fit cap in the bottom of a T can fall out. It also can get stuck in, due to sticky creosote or corrosion. Design changes for a more secure closure and easier removal would be helpful.

Lightning Protection

If lightning protection is recommended in your area, because of the number and severity of thunderstorms, all chimneys, including those serving wood burning equipment should be protected.

A lightning protection system doesn't prevent lightning from striking but it does provide a safe path to the ground for the electric current when lightning strikes. When an unprotected, non-metallic building is hit, the current must pass through non-conducting materials such as masonry and wood. The high electrical resistance of these materials causes so much electrical heating that fires or explosions can result. The same amount of current can flow through a metal building or a lightning protection system without a buildup of heat, and hence without damage.

Since lightning tends to strike high points on a structure, chimneys are likely targets. Masonry chimneys require the same kind of protection as does the whole roof. When adding a masonry chimney to a house with lightning protection, an additional air terminal (metal rod sticking up into the air at or near the chimney) and connecting cable should be installed. If the house has no protection, an air terminal should not be added to a chimney alone. Either no system is necessary, because there are not enough severe lightning storms to constitute a risk, or the whole house should be protected.

Metal chimneys are good enough electrical conductors to be part of a lightning protection system. They can carry the electrical current without being damaged or being a hazard to the structure. If they are not grounded, at some point the current will jump away from the chimney or stove, to or through the house structure, and this is dangerous. For the most effective and durable lightning protection systems local building codes should be followed. For temporary protection, heavy copper (¼-inch diameter solid or AWG 2 stranded) or aluminum (⅓-inch diameter solid or AWG 0 stranded) uninsulated cable should be wrapped twice around the metal chimney at any convenient location and grounded, by connecting to the water main entering the house (if there is one and if it is metal, not plastic pipe) (Figures 2–7 and 2–8).

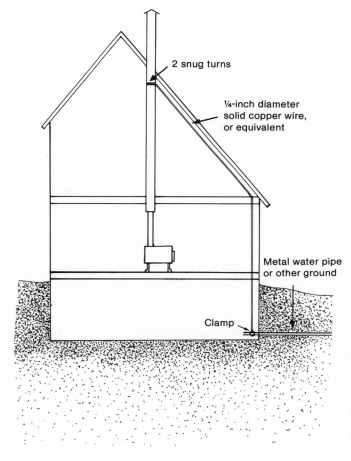

2 snug turns

¼-inch diameter solid copper wire, or equivalent

Metal water pipe or other ground

Clamp

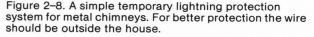

Figure 2–8. A simple temporary lightning protection system for metal chimneys. For better protection the wire should be outside the house.

NEW CHIMNEYS

Stovepipe is *not* a safe material for a chimney, although it can be used for the chimney connector. When used outside a house it corrodes very quickly, typically in one heating season, even if galvanized. It lacks mechanical strength, and because it can get so hot, it requires a minimum of 18 inches of clearance from any combustible material, according to the National Fire Protection Association (NFPA). NFPA and most building codes forbid its use as a residential chimney. In fact no single-wall metal chimney, even if heavy gauge, is allowed inside one- and two-family residences.

There are two types of potentially safe chimneys—prefabricated or factory-built metal chimneys, and masonry chimneys. I say "potentially safe" because no common chimney type is totally safe if improperly installed or if abused in use.

New Factory-Built Chimneys

The factory-built metal chimneys are easier to install and may be less expensive especially where masons' rates are high. They have two, three, or four metal walls with either air or insulation in between (Figures 2–11 and 2–12). The innermost layer is generally stainless steel, to withstand the high temperatures and corrosive environment. The insulation used in most chimneys is a powder of amorphous silica plus a little fiberglass; it does not contain asbestos. Here are four important points concerning factory-built chimneys.

1. Only "Class A" or "All-Fuel" or "Solid Fuel" chimneys, intended for use with wood heaters, should be used. There are other kinds of similar-looking metal chimneys and vents that can easily *melt* if used with wood stoves. They have aluminum inner liners and are intended primarily for use with gas appliances.

2. The chimney should have a UL (Underwriters Laboratories, Inc.) listing, or be approved or listed by another appropriate organization. Each chimney section and support part should have a label saying "UL Listed." This means that the chimney model has passed a safety test.

3. Prefabricated chimneys must be installed according to the manufacturer's instructions.

Figure 2–9. Government standards for wood-heating installations have changed over the years. The illustration is part of a 1799 engraving showing the west end of Independence Hall where the Continental Congress met and the Declaration of Independence was signed. Note at left the two stovepipes protruding out the windows. By today's standards, the installation was probably illegal on the following counts: material, wall pass-through, clearances, mechanical support, and height. (Illustration courtesy of Independence National Historical Park Collection. I am grateful to Prof. Samuel Edgerton of Boston University for bringing this illustration to my attention.)

Figure 2–10. Examples of stovepipe chimneys. All building codes prohibit stovepipe chimneys for residences. The middle example has proper clearances and mechanical support but is still illegal where codes must be followed. Stovepipe chimneys are unsafe principally because of their short life expectancy, mechanical frailty, susceptibility to creosote accumulation, and high exterior surface temperature.

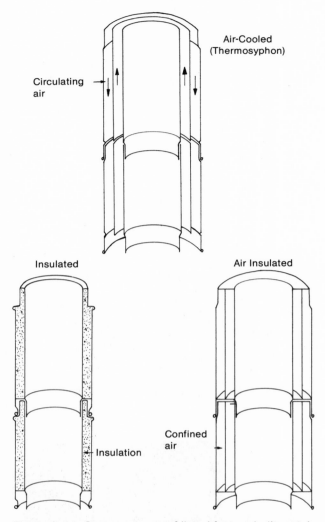

Figure 2–11. Common types of listed factory-built metal chimneys. All are essentially equally safe when installed in accordance with instructions and when properly maintained—that is, cleaned frequently enough so that intense chimney fires do not occur. However the necessary cleaning frequency may depend on chimney type.

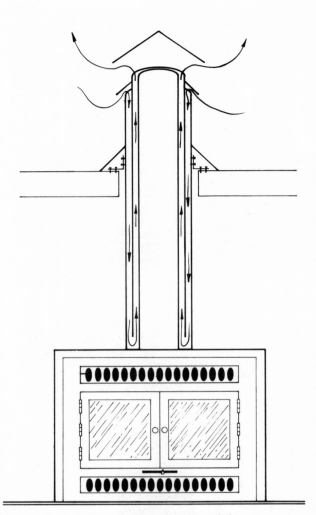

Figure 2–12. The operation of an air-cooled or *thermosyphon* chimney. The hotter inner air layer tends to rise, causing the air circulation pattern shown. The objective is to keep the outer surface of the chimney at a safe temperature.

(Underwriters Laboratories checks the adequacy of instructions as part of its listing procedure.) Insulated chimneys are not cool on the outside; *most prefabricated chimneys require a minimum of 2" clearance to any combustible material* (Figure 2–13). All other details of the instructions must also be followed carefully (Figure 2–14); otherwise the UL listing may be meaningless since Underwriters Laboratories only tests the chimney installed as specified in the instructions. Mixing pieces and components from different manufacturers does not usually work mechanically since they are not usually compatible, and even with an apparently good fit, the result may not be safe.

4. The chimney must be kept relatively clear of creosote and soot deposits. UL testing of chimneys includes simulation of a chimney fire (the burning of deposits inside the chimney) by using flue gas temperatures of 1700° F. for 10 minutes, following a flue gas temperature of 1000° F. for as long as necessary to establish equilibrium temperatures of the chimney and surrounding structures. Chimney fires can be even more severe than this. Thus chimneys must be inspected frequently and cleaned when needed to prevent intense chimney fires. If the build-up of soot and creosote exceeds a quarter-inch in thickness, I recommend cleaning the chimney.

Figure 2-13. Factory-built metal chimney installations without adequate clearance to combustibles.

Figure 2-14. A consequence of inadequate mechanical support of a factory-built chimney. This chimney now has a large opening in it between two sections just below the roof overhang. Had this tipping of the chimney occurred while the chimney was in use, a house fire would have been quite possible. Factory-built chimneys extending more than about 4 feet should have extra support, such as with two roof-attached rods connecting to a strap around the chimney.

New Masonry Chimneys

Masonry chimneys require a foundation or footing well below the frost line. Foundations are typically reinforced concrete, about 10 inches thick, and extend about 12 inches beyond the chimney sides. Local code requirements vary somewhat.

The three most common masonry chimney materials are bricks, concrete blocks of various shapes, and stone (Figure 2-16). According to NFPA, the bricks or concrete blocks must be at least 4 inches thick. (Bricks and other standard masonry units are roughly 3½ inches wide, not 4. Thus most masonry chimneys are in technical violation of NFPA standards. The standard probably should be changed to reflect common and accepted practice.)

NFPA standards state that stone walls must be at least 12 inches thick. Natural stone typically has a higher thermal conductivity than concrete or brick, and also tends to come in irregular shapes; thus thicker walls are sometimes needed to be as strong and to prevent dangerously hot chimney surface temperatures. (See Table A2-3 in Appendix 2 for thermal resistances of various masonry materials.) However 12 inches may be unnecessarily conservative.

Concrete block chimneys are the easiest and least expensive to build, but they have a reputation for being less safe than brick or stone chimneys. Part of the problem may be that a larger proportion of them are built by homeowners, not professionals. Common mistakes are inadequate clearances, inadequate foundations, and even the absence of liners.

But there may also be more subtle safety problems. Not any kind of concrete is suitable—low thermal expansion is desirable. Concrete chimney blocks with integral liners are highly susceptible to cracking. Many concretes are more susceptible to heat stress and corrosion from flue gases and creosote than are ceramic materials. The relatively porous structure of typical concrete blocks may result in higher weathering rates. Adequate curing of concrete before subjecting it to thermal stress may be important.

Some concrete ring blocks have significantly thinner walls than NFPA's recommended 4 inches. Others are too small on the inside; the liner fits so snugly that little insulating air space is left and the chimney may crack when the liner expands as it gets hot.

19

Figure 2–15. Masonry chimneys with foundation problems.

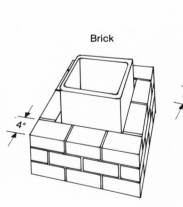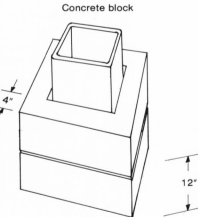

Brick

Concrete block

Stone

Figure 2–16. Common types of masonry chimneys. The space between the liner and the bricks or concrete blocks should not be filled with mortar; stone chimneys should be constructed with a similar air gap, roughly ½ to 1 inch.

Despite all these potential problems, I believe concrete chimneys are reasonably safe when built properly—with a liner and with adequate clearances—and when kept clean enough to avoid intense chimney fires. Interior locations are preferable because there is less weathering and less creosote build-up. A concrete sealer applied to exterior chimneys can retard weathering.

A tight-fitting metal cleanout door should be installed near the base of the chimney. Its frame should be set without air leakage into the chimney wall. A fire-clay lining not less than ⅝-inch thick should be installed ahead of the masonry units as the chimney is built up. It should start at the bottom and extend the full height of the chimney, extending at least a few inches above the highest masonry units.

Refractory mortar, not ordinary mortar, should be used between the liner sections. The joints should be as tight as possible and finished to leave a smooth surface inside the liner.

The liner should be separate from the chimney wall. The roughly half-inch space between the liner and chimney wall should not be filled with mortar, although enough mortar should be used at each joint to squeeze out and touch the chimney wall and thus hold the liner in position in the middle. This lack of rigid connection between liner and chimney allows the liner to expand and contract as it heats and cools without stressing and cracking the chimney wall.

Filling the space between liner and chimney with a high temperature insulation such as perlite, vermiculite, or ceramic wool would improve draft and reduce creosote accumulation. Exterior temperatures of the chimney would also be reduced. This variation on masonry chimney construction has not been tested in laboratories or, more importantly, been used much in houses. Thus its actual life performance and durability are not known. The idea is promising.

Adding insulation in an *interior* exposed masonry chimney reduces the house-heating effect of the chimney, thus lowering the energy efficiency of the system. Also mechanical support of the liner is critical; creating a larger than normal space between liner and chimney wall in order to get more insulation in would be dangerous unless extra support for the liner were provided. Settling, sagging, wetting, and freezing of the insulation should also be anticipated. Finally, the higher temperatures of the liner will result in additional thermal expansion that could cause mechanical problems.

For those who want to experiment with an insulated masonry chimney, I would suggest using about 2 inches of perlite or vermiculite for the insulation, supporting the liner every foot and the insulation about every 2 feet (Figure 2–19), and perhaps experimenting with a slightly flexible (sheet metal?) drip cap or flashing at the chimney top both to keep moisture out of the insulation and to allow for the different thermal expansion of the liner and the masonry.

If two flues are being built into the same chimney, the joints of the two liners should be staggered by at least 7 inches (Figure 2–20). If more than two liners are being installed, masonry *wythes* or partitions are required by NFPA to separate the flues into groups of not more than two. Many masons build wythes between all flues in a chimney. In this case the liner joints need not be staggered. The partition should be about 4 inches thick (a brick width) and be secured into the chimney walls.

Figure 2–17. Sewer pipe is not designed for use as a chimney or even as a chimney liner. The clay cracks at high temperatures.

21

Figure 2–18. A finished chimney and a chimney under construction at Plimoth Plantation, a reconstruction of the first Pilgrim community at Plymouth, Mass. The chimneys were made of wood and mud. Wood beams were even placed inside the flues to support cooking pots over the fire. Standards have changed, but so has usage. These chimneys served huge fireplaces with relatively small cooking fires. The vast quantity of air that went up the chimney along with the smoke lowered the temperature to lukewarm at most. Under these conditions wood chimneys are not quite as outrageous as they might seem.

Care must be taken to assure airtight separation of the flue passages at the *bottom* of the multi-flue chimney. A separate cleanout door is required for each flue. The bottom of the first section of each liner should be installed so that each flue and its own cleanout door are sealed off from all others. A potential problem is that a stronger draft in one flue can result in reduced draft or even reversed downwards flow in the adjacent flue if there is a connection between the two flues at their bottoms.

Clearance Around Masonry Chimneys

This book recommends following the clearances recommended by the Canadian Heating, Ventilating, and Air-Conditioning Code (1977) rather than those in NFPA 211–1977.

The Canadian code seems more logical, more complete and more conservative than NFPA 211. According to the latter, wood beams, joists, and studs require a 2-inch clearance, except where the chimney wall is 8 inches thick. However wood lathing and furring require no clearance when within 1½ inches of the corner of the chimney even if the chimney wall is only 4 inches thick. Thus the wood can be within about 4⅝ inches of the liner. Exterior chimneys require *no* clearance from the building. Nothing is said about clearances for flooring, trim, and paneling.

Corners of chimneys are generally cooler than the centers of the sides. This may be the rationale for permitting furring near the corners of standard 4-inch wall masonry chimneys. But then why not also allow beams, joists, and studs to contact chimney corners? Perhaps they dissipate heat less effectively. And if clearances of any type are required for interior chimneys, why is none required for exterior chimneys? Perhaps the extra cooling of exterior chimneys due to colder outdoor temperatures is significant. Perhaps also house fires starting on the exterior of a house are slightly less dangerous because they spread through the rest of the structure more slowly.

The Canadian code requires 2 inches of clearance between combustible materials and the outside of a masonry chimney. There are three exceptions to this:

1. Exterior chimneys need have only ½-inch clearance from the combustible wall next to it. NFPA requires no clearance here.

2. Wood or other combustible flooring must be kept at least ½-inch short of a masonry chimney.

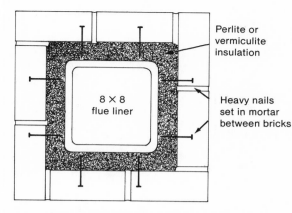

Figure 2–19. A suggested experimental insulated masonry chimney design. The nails keep the liner centered. Two sets should be used for each 2-foot liner section. Strips of sheet metal should be laid across the nails about every 2 feet to prevent settling of the insulation. Or mason's sheet metal ties can be used just below each junction between liners to support the mortar overrun, which then supports the liner and keeps the insulation from settling. (Riteway Manufacturing Co. has been suggesting this general type of chimney for many years.)

3. Wood trim must have at least ⅛-inch of asbestos or other noncombustible material between it and the chimney. I expect sheet metal would be a reasonable alternative to asbestos, particularly if the assembly is loose, leaving small air gaps.

Thus nowhere should wood be in direct contact with a masonry chimney.

NFPA permits wood girders to be supported on a corbeled shelf of a masonry chimney if the chimney wall is at least 8 inches thick at that point. This is really a "grandfather" provision since in contemporary construction chimneys are not and should not be used to support the building.

Where the chimney passes through floors or ceilings, the 2-inch clear space between the chimney and the surrounding wood structure should be "firestopped" with non-combustible material no thicker than 1 inch. The intent is to slow the spread of a fire. A sheet-steel metal flange or frame between the chimney and the surrounding wood framing is adequate. Bringing ceiling Sheetrock right up to the chimney will also do.

The space between the chimney and nearby joists should not be filled with insulation because doing so can result in *higher* temperatures at the wood. Convection in the relatively open air space usually does a better job keeping the wood cool than does insulation in this situation. Obviously,

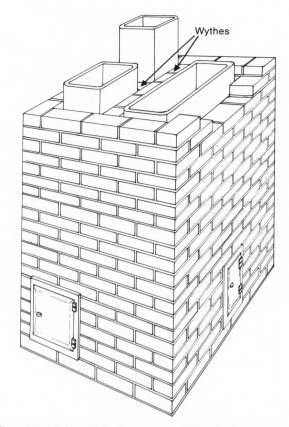

Figure 2–20. A multi-flue masonry chimney. Many masons put *wythes* (interior walls) between the flues, but NFPA allows pairs of flue liners to be directly adjacent to each other if their joints are staggered by at least 7 inches. Each flue should have its own cleanout door and should be isolated from the other flues at the bottom of the chimney so the drafts are independent.

combustible insulation should never be placed against chimneys. Cellulose insulation blown into an attic must be pulled away from chimneys and from recessed lighting fixtures in the ceiling below.

Where clearances are minimal or when additional protection is desired, sheet metal between the chimney and the house will help. Ideally the sheet metal should be mounted half-way between the chimney and the wood needing protection, and in such a way that air can circulate on both sides of the metal, for it is principally the circulating air that helps keep the wood cool. If the situation makes this difficult, the sheet metal will still give some additional protection whether it is mounted on the chimney or on the wood; or even just squeezed between the two. Shiny metals such as aluminum or galvanized steel are best. Aluminum flashing is easy to obtain and work with.

23

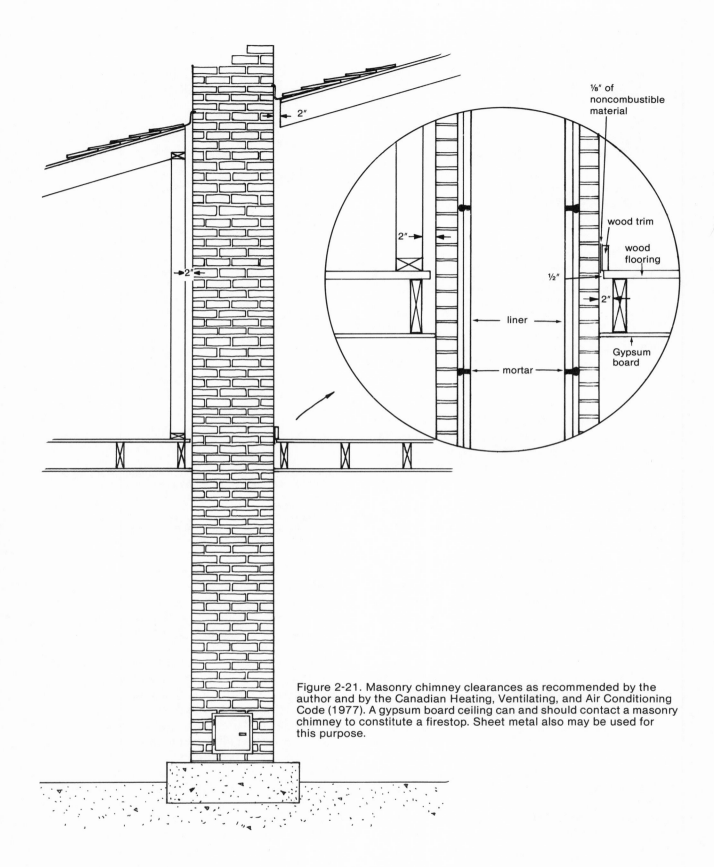

2"

⅛" of
noncombustible
material

2"

wood trim

wood
flooring

2"

½"

liner

Gypsum
board

mortar

Figure 2-21. Masonry chimney clearances as recommended by the author and by the Canadian Heating, Ventilating, and Air Conditioning Code (1977). A gypsum board ceiling can and should contact a masonry chimney to constitute a firestop. Sheet metal also may be used for this purpose.

24

Figure 2–22. An exterior masonry chimney *with* clearance between it and the house, a good feature for chimneys serving solid-fuel heaters.

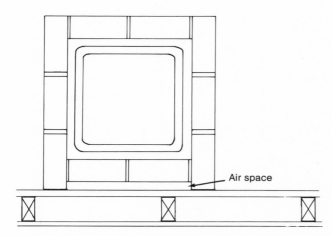

Air space

Figure 2–23. A suggested design for an exterior masonry chimney built next to combustible walls. NFPA allows exterior chimneys to touch the house; the Canadian building code does not. The illustrated design is a compromise. Ventilation of the air space is probably unnecessary.

EXISTING OR USED CHIMNEYS

New chimneys are expensive—often in the range of $10 to $20 per foot of height. This, plus the effort and time involved in building or installing new chimneys, makes it attractive to use an existing chimney if at all possible.

One should be extremely cautious about connecting a wood burner to just any chimney that happens to be available. There are four reasons for concern:

1. Not all chimney types are safe for solid-fuel heaters. This is particularly true of some types of prefabricated metal chimneys.

2. Old chimneys may have deteriorated, even if unused.

3. Even if still in good-as-new condition, many old chimneys do not meet today's safety standards for wood heater chimneys.

4. Some other appliance, such as a furnace or water heater, may already be hooked to the flue. In general, multiple use of flues is not advisable.

Used Metal Chimneys

Be sure any factory-built chimney was designed for use by wood heaters. Look for labels saying "All Fuel" or "Class A" or "Solid Fuel." Chimneys designed for gas appliances can melt if used for wood-burning appliances. If in doubt, ask a building inspector or fire chief.

Metal chimneys should not show signs of significant corrosion or rust, and they should be mechanically sound. Joints between sections should be secure, and the chimney should be supported about every 10 feet and within 5 feet of the top with non-combustible brackets, spacers, or rods.

And of course the chimney should be clean. If the chimney has mechanical shortcomings, they may be quite apparent during cleaning.

Used Masonry Chimneys

For better or worse, safety is a matter of degree. No installation is absolutely safe against all imaginable circumstances. Old masonry chimneys are often in a gray area. It is difficult to assess

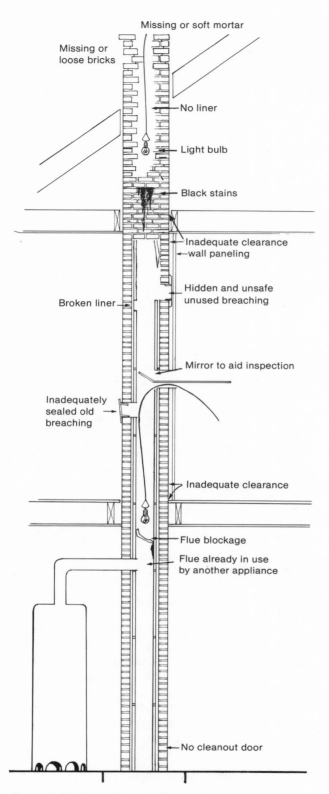

Missing or soft mortar

Missing or
loose bricks

No liner

Light bulb

Black stains

Inadequate clearance
wall paneling

Hidden and unsafe
unused breaching

Broken liner

Mirror to aid inspection

Inadequately
sealed old
breaching

Inadequate clearance

Flue blockage

Flue already in use
by another appliance

No cleanout door

Figure 2–24. Possible problems to look for when examining a masonry chimney. If a light bulb is used, glare can be reduced by using a very small aluminum foil shade on the portion of the bulb facing the viewer.

accurately their condition, since they are not often wholly accessible for inspection. Even if their exact condition were known, there is no simple answer to the question, "How safe is safe enough?"

I suggest cleaning the chimney *before* deciding whether to use it. While cleaning you may discover mechanical weaknesses such as loose bricks, soft mortar, cracked liners, or unused breachings which might not be detected in a strictly visual examination.

Examine the exterior of the chimney over its entire length if possible, and this includes in the basement and attic where applicable, and above the roof.

The chimney interior should be examined in as many locations as possible. Most chimneys have at least three holes into them—a cleanout door near the base, a "breaching" where an appliance was or is connected, and the top of the chimney. Use a hand-held or stick-mounted small mirror to look up or down a chimney interior, and a light bulb on a long cord for checking down into a chimney from its top.

In examining the chimney inside and out check for:

1. *Black or dark brown stains or drips* on the chimney's exterior. These are either creosote or evidence of rain water having gotten into the chimney and then leaked out. In either case the chimney has, or at least once had, defects.

2. *Soft mortar.* It is very difficult to drive a nail into good mortar.

3. *Loose, missing, or cracked masonry units* (bricks, blocks, stones). If the chimney has any of these defects, consider either not using it, or adding a good metal liner. Why not repair it? The failings are fundamental, and only rarely is a chimney accessible enough to be able to locate and repair all the defects. If you are fairly sure the only such failings are in the chimney above the roof, repair is possible. All loose mortar can be scraped out and renewed.

4. *An intact liner.* All masonry chimneys should have fireclay liners. They help to keep chimneys "smoke-tight," and to prevent overheating and cracking of the brick, block, or stone chimney wall, particularly during very hot fires or chimney fires.

Lining an existing chimney is difficult but possible.

Some people have managed to install ordinary tile liners by putting mortar or furnace cement on

Figure 2–25. Masonry chimneys needing work to make them safe for use.

the tops and lowering the tiles into the chimney. Clearly this can be done only in a straight chimney, and even then leaks between lengths are likely. Still, some additional protection will be gained.

A safer way to get a tile liner into an existing masonry chimney is to knock a few large temporary holes in the chimney so that careful placement and mortaring of the liner are possible. This requires considerable work, but you know what you have when you're done.

Stovepipe is not a very durable liner. Although it offers some protection while intact, it is likely to rust out in a year or so. Because the pipe is hidden in the chimney, the homeowner may not realize its deteriorating condition. Or the replacement job may be postponed indefinitely because it can be difficult and messy. A more permanent liner is highly preferable.

A high-temperature-tolerant stainless steel liner functions well. Some metal shops can fabricate stainless steel "stovepipe," and it is now beginning to be manufactured (see Appendix 5). Joints should be screwed together and the system made mechanically secure so that chimney cleaning inside the pipe is practical. A wire brush

is probably the best cleaning device to use and will not snag on the protruding sheet metal screws.

Another good type of liner for retrofitting chimneys is made of enamel-coated heavy gauge steel (Appendix 5). In England fireclay pipes with socketed joints, like stovepipe joints, are available. Both these liners are safer as retrofits than ordinary straight American liners because of the way the joints fit together providing a moderately good seal, some mechanical security, and retention of all dripping creosote inside the liner, and all of this without use of mortar.

No type of readily available liner can be easily retrofitted into a chimney that does not run straight up. Flexible stainless steel pipe could work but would be very expensive.

A certainly safe liner to add to an existing chimney is regular factory-built chimney sections. The chimney must be straight and sufficiently large to end up with adequately large inner flue passage. The outer diameter of factory-built chimneys is typically 3 to 6 inches larger than the inner or flue diameter.

5. *Obstructions.* Broken liners, fallen or loose bricks, animal nests, improperly installed chim-

27

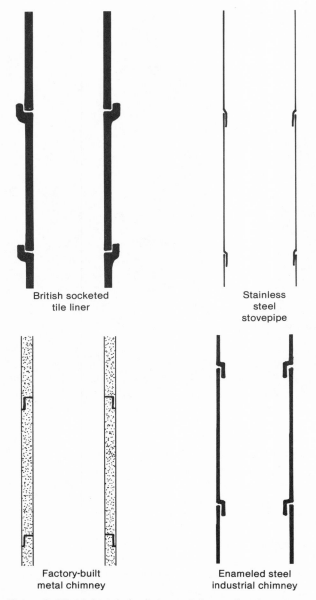

British socketed tile liner

Stainless steel stovepipe

Factory-built metal chimney

Enameled steel industrial chimney

Figure 2–26. Materials for lining existing unlined masonry chimneys. Heavy steel pipe such as well casing may also be suitable although it will eventually corrode. Clay sewer pipe, although socketed and noncombustible, should not be used. The heat will crack and degrade the clay.

ney connectors and creosote accumulations can all cause serious blockages.

6. *Unused breachings.* Many old chimneys apparently served to vent small stoves in every room they passed through. Five or more breachings into a single-flue chimney are not uncommon. It is important to locate and safely seal the unused breachings. Examining the ex-

terior of the chimney may reveal locations of some openings, often covered with a thin metal "pie-plate" cap, but others may be covered by plaster or wall paneling. Try to locate these by examining the chimney's interior, or by tapping the exterior and noting the difference in the sound. This can be difficult and some may be missed; this is a risk in using old chimneys, since some of the breachings may not be safely sealed. It is especially important to check remodeled portions of a house. A number of house fires have been started by chimney fires igniting paneling over a pie-plate cover.

To seal an unused opening into a chimney, use masonry materials such as bricks and mortar to make the former opening as solid and tight as the rest of the chimney. The standard type of pie-plate cap is not safe because it is not securely airtight, has virtually no insulating value, and can be blown out by a pressure surge inside the chimney.

7. *Cleanout door.* Although not themselves directly related to safety, cleanout doors contribute indirectly by making it easier to clean chimneys. If a masonry chimney lacks a cleanout door, one can be added. The original masonry must be chiseled or drilled out (use eye protection) and the door mortared in.

8. *Clearances.* Masonry chimneys can get *hot*—so hot, in fact, that not only is it considered unsafe (in Canada at least) to have any direct contact between the chimney and wood or any other combustible material, but most combustible material should be at least 2 inches away from the chimney (see Figure 2–21). Common construction practices and safety standards have changed over the years. Many if not most old chimneys fail the clearance test; using them involves extra risk and requires extra precautions to avoid chimney fires and sustained very hot normal fires. As discussed previously, placing sheet metal, such as aluminum flashing, between the chimney and nearby wood gives some protection, particularly if there is a small airspace on both sides of the metal.

Smoke Test

Even a careful inspection of a masonry chimney may not reveal all defects, particularly cracks and holes through the mortar, and inadequately sealed breachings. A smoke test of the chimney can help (Figure 2–27). A *small* smoky fire or a non-toxic smoke bomb is placed in the appliance or the base of the chimney. A non-self-starting chimney may

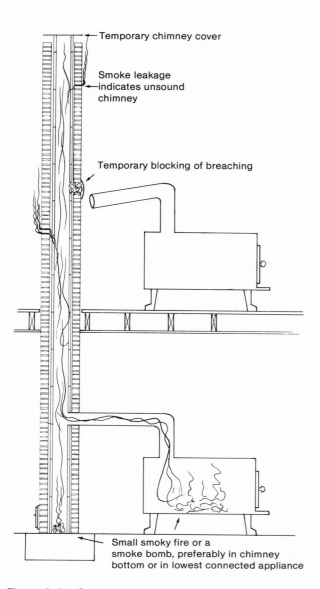

Temporary chimney cover

Smoke leakage indicates unsound chimney

Temporary blocking of breaching

Small smoky fire or a smoke bomb, preferably in chimney bottom or in lowest connected appliance

Figure 2–27. Smoke test for tightness of chimneys. Any extra appliances venting into the same flue, such as the upper stove in the illustration, should be disconnected and the breachings into the chimney sealed during the test; otherwise annoying and confusing amounts of smoke may enter the house through leaks in the appliance and its connector.

have to be warmed up a little with a small fire to get the smoke from the bomb or smoky fire to rise and fill the chimney. The chimney top is then blocked so that positive pressure is created everywhere inside the chimney. Smoke seen coming through the chimney wall will point out leaks needing repair. If the number of smoke wisps is large the chimney may be in such poor condition that no amount of repair will make it safe.

But if it was good enough for granddad, why isn't it good enough for me?

Here are four reasons:

1. Granddads may have been more knowledgeable and hence more cautious and safer in how they operated and maintained their heating systems.

2. We may be subjecting chimneys to more severe conditions. For instance, flue gas temperatures in chimneys serving open masonry fireplaces cannot normally get very high because of all the diluting room air which escapes up the chimney. Adding glass doors or some fireplace insert devices, or installing large stoves in fireplaces *can* result in much higher temperatures than the designer or builder expected.

3. Our safety standards are higher. Perhaps statistically granddad *did* burn his house down every so often—more often than we would find acceptable today.

4. Parts of our houses may be more flammable—such as some synthetic carpets, drapes, and floor finishes. Also fumes from some plastics, insulating foams, and furniture padding are more lethal than smoke from natural materials.

MULTIPLE USE OF SINGLE FLUE

Is it safe to attach a wood heater such as a stove, furnace, or boiler to a chimney flue that is also used by another appliance such as a fossil fuel furnace or water heater, a fireplace, or another stove?

Since new chimneys are so expensive, often costing more than a stove, it is tempting to use an existing flue even if it is serving another appliance. A few states have prohibited combining wood and oil or wood and gas in a common flue, but many allow it. The National Fire Protection Association says, "Don't do it, but if you do, here's how."[2]

The distinction between a chimney and a flue is important. Masonry chimneys today are often built to serve more than one appliance by having separate flues or tile-liners within the overall

2. Freely paraphrased from *Using Coal and Wood Stoves Safely*, NFPA HS-10, 1978, National Fire Protection Association, Boston, Mass.

masonry structure (Figure 2–20). Each flue or liner is an isolated independent channel. Thus it is perfectly safe to attach a number of appliances to the same chimney if only one is attached to each flue, and each flue is of appropriate size for the individual appliance. When building a new house with one or more masonry chimneys, it is wise to install enough separate flues to handle each anticipated appliance.

Many older houses have only one chimney and it has only one flue. The owner is then faced with the question of whether to use the same flue for, as an example, a wood stove and a furnace.

There are seven areas of concern:

1. According to most building codes, fireplaces and fireplace stoves (those that can be operated with the doors open, such as Franklin stoves) should not share a flue with any other appliance—wood or fossil fuel.

Because fireplaces and fireplace stoves can be left open, sparks or burning creosote from another appliance may come out into the room through them. Thus some believe that a stove should be connected to a fireplace flue only if the fireplace, or its flue, below the opening for the stove, is well sealed. Most building codes require a "permanent" brck and mortar or sheet metal seal over the fireplace opening or in the throat or flue, even when the stove is a floor or two above the fireplace.

I think that danger related to sparks coming out of open-type appliances due to use of other appliances sharing the same flue is small—and in particular, small compared to the use of the open wood heater itself, which codes do not prohibit. Natural drafts would usually impede the movement of sparks down into and out the open appliance. Even if glowing or burning material were to fall into a fireplace or fireplace stove, it would have to travel outwards a considerable distance to clear the hearth extension or floor protector in order to land on a combustible part of the house. And finally, if dampers are left closed and if spark screens or doors are in place over the opening when the units are not in use, sparks and burning creosote are unlikely to get out.

Another more serious problem with open wood burners is their very large air consumption. Both the quantity of air flow up the chimney and its relatively cool temperature may deprive other appliances connected to the same flue of their recommended drafts. As explained below, this is of more concern when the "other" appliance burns oil or gas than when both appliances burn wood. Some chimneys will always provide adequate draft, but as a general rule, I do not recommend venting a fossil fuel appliance and a fireplace stove (or fireplace) in the same flue.

2. A common flue must have adequate capacity to remove safely the products of combustion from all connected appliances and it must generate enough draft for safe combustion in each appliance.

If the draft at an oil or gas burner is inadequate, the fuel will not burn completely and dangerous fumes will be generated. If the chimney capacity is inadequate, the fumes or flue gases will not vent out of the house properly.

Although usually when a wood heater is in use the burner in the central heating system will be off, it is absurd to presume this will always be the case. The flue must be designed to handle both appliances under *all* conditions.

Connection of several appliances to one flue is common in commercial and industrial buildings. In homes, furnaces and water heaters often share the same flue. In these cases the only real concern is the adequacy of capacity and draft.

There are two common criteria for determining the adequacy of a flue to handle more than one appliance. One simply requires that the common flue's cross-sectional area be at least as large as the sum of the minimum required flue areas for each appliance alone. (Lacking explicit information from the manufacturer, it is reasonable to take the flue collar area to be the minimum required flue area.) This simple method implicitly recognizes the dominant importance of area in chimney sizing. Some codes require the common flue's area to be the sum of the area required by the largest appliance plus 50 percent of each area required by the other appliances.

Another approach to common-flue sizing requires knowing the *input* ratings of each appliance and takes into account the effect of chimney height on capacity (Table 2–2).

3. Because creosote and soot deposits in chimneys are virtually inevitable when burning wood or coal, a flue may become so choked that its capacity is no longer adequate. If a flue serves only one closed-type wood stove, a decrease in capacity may be inconvenient but is not dangerous. The decreased draft means less air enters the stove and the combustion rate is slowed—the stove will not get as hot, and each fuel loading will take longer to burn. However, in gas and oil appliances the fuel is injected at a constant rate. If the air supplied is inadequate for combustion, danger-

Table 2-2. Chimney Capacity

Height of Flue (Feet)	AREA OF FLUE (SQUARE INCHES)					
	19	28	38	50	78	113
	Combined Appliance Input Rating (Thousands of Btu per Hour)					
6	45	71	102	142	245	358
8	52	81	118	162	277	405
10	56	89	129	175	300	450
15	66	105	150	210	360	540
20	74	120	170	240	415	640
30	83	135	195	275	490	740

This table is adapted from Table 17 in chapter 26 of the 1975 Equipment Volume of the *ASHRAE Handbook and Product Directory;* essentially the same table appears in the National Fire Protection Association booklet, *Using Coal and Wood Stoves Safely.*

Maximum input capacity (in thousands of Btu per hour) of chimneys used by more than one appliance. This table should not be used for solid fuel appliances intended to be operated with an open fire, such as fireplaces and fireplace stoves. Input ratings of wood stoves are about twice their rated heat output, since their efficiencies are about 50 percent. Input ratings of other appliances, such as furnaces and water heaters, are usually indicated on them, but again, twice the rated output is a reasonable assumption for wood burning equipment.

Although this table was not originally designed for solid fuel appliances, the numbers are probably roughly reasonable for this case. As an example of the use of this table, consider the following situation: suppose an oil furnace with an input rating of 120,000 Btu per hour were vented through a 20-foot masonry chimney with an 8″ x 8″ (nominal outside dimensions) tile liner. Would this chimney have the capacity to be used also by a wood stove with a rated power output of 40,000 Btu per hour?

Use of the table requires power input ratings. Since stoves, furnaces and boilers are roughly 50 percent efficient, the input rating is about twice the output, or 80,000 Btu per hour in this case. Thus the combined input ratings of both appliances would be 120,000 + 80,000 = 200,000 Btu per hour. The inside area of an 8- x 8-inch tile liner is about 46 square inches. From the table it can be estimated that the capacity of a 20-foot chimney with an area of 46 square inches would be about 215,000 Btu per hour, enough to handle the furnace and stove. (For purposes of using this table, chimney heights should be measured upwards from the upper breaching into the chimney.)

ous amounts of unburned or partially burned gases will be created. With the chimney being partially blocked, gases may spill out of the appliance into the house.

4. The accumulation in the chimney bottom of fallen creosote and soot may block the appliances venting into the lower portions of the chimney. This is especially likely when the cleanout door is located just below the lowest breaching, but of course it can happen in any installation if maintenance is inadequate. This problem is not restricted to multiple-use flues, but wood heaters usually result in so much creosote accumulation that the situation can quickly become more serious when a wood heater is added to a flue.

5. One of the best ways to control a chimney fire is lost if the flue serves more than one appliance. When a single airtight stove is connected to its own airtight flue, merely closing the air inlet on the stove will deprive the flue of oxygen and thus suffocate the chimney fire. When an oil or gas appliance or even a second stove is connected to the same flue, stopping all air flow into the flue becomes much more difficult.

6. Under rare circumstances, small explosions inside oil and gas appliances may occur. Gas and oil burners are equipped with primary safety controls which stop the flow of fuel if it is not ignited or if the flame goes out. But there is a slight and necessary delay, so that a few seconds' worth of unburned gas or oil spray may enter the appliance. Subsequent ignition may lead to a small explosion in the appliance. Alternatively, the combustible gas may be pulled up the chimney. If at this time a wood heater is venting into the same flue, it is conceivable that the gas or oil "fog" could be ignited, resulting in a small explosion in the chimney.

These explosions do not usually result in serious permanent damage to the appliances or chimneys. But having a stove hooked up to the same flue could conceivably lead to a dangerous situation if the stovepipe connector is blown out of the chimney so that the wood heater is venting into the house, or if the pressure surge of the explosion forces sparks and other burning material out of the wood heater's open front or air inlet damper. The likelihood of these occurrences is not well established, but such explosions have happened.

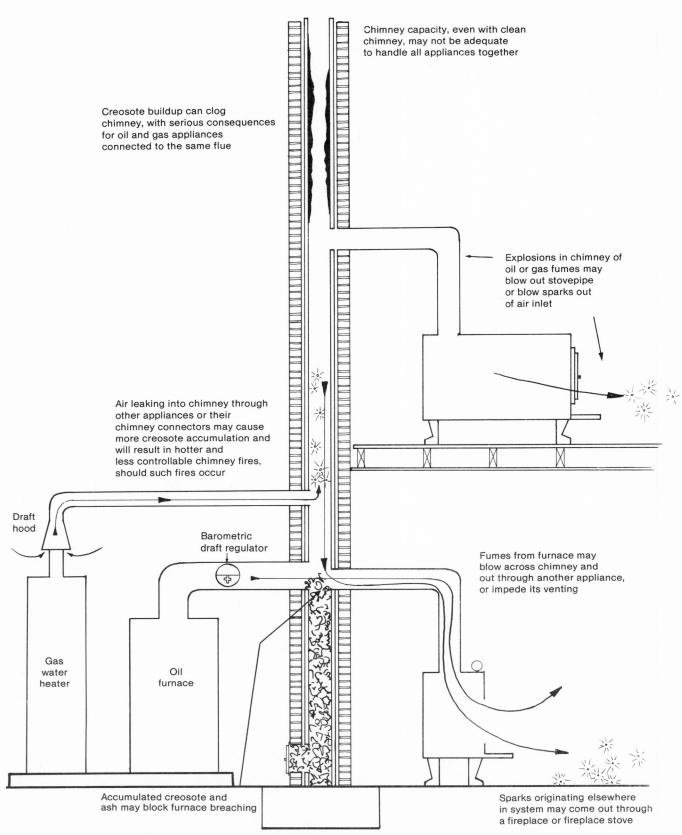

Chimney capacity, even with clean chimney, may not be adequate to handle all appliances together

Creosote buildup can clog chimney, with serious consequences for oil and gas appliances connected to the same flue

Explosions in chimney of oil or gas fumes may blow out stovepipe or blow sparks out of air inlet

Air leaking into chimney through other appliances or their chimney connectors may cause more creosote accumulation and will result in hotter and less controllable chimney fires, should such fires occur

Draft hood

Barometric draft regulator

Fumes from furnace may blow across chimney and out through another appliance, or impede its venting

Gas water heater

Oil furnace

Accumulated creosote and ash may block furnace breaching

Sparks originating elsewhere in system may come out through a fireplace or fireplace stove

Figure 2–28. Possible consequences of connecting more than one appliance to the same flue.

32

7. Under some circumstances creosote accumulation is increased in shared flues. All appliances let some air into their chimneys even when the appliance is not in use. Considerable air is allowed into flues through draft hoods on gas appliances and through barometric draft regulators used with oil appliances. This air leakage into the chimney through draft regulators is important for the safe and efficient operation of the fossil fuel appliances, but the cooling effect in the chimney can cause more creosote accumulation. This effect is probably largest when the fossil fuel appliance rarely comes on, such as when a wood stove, furnace, or boiler has taken over almost all of the heating of a house.

Multiple use of a single flue can also result in *less* creosote accumulation. Either the fossil fuel appliance or the wood heater is likely to be in use most of the time. Thus a shared chimney can be warmer than separate flues would be. When this is true less creosote accumulates. This is probably the dominant effect when the wood appliance is being used no more than a small portion of each day. This possible reduction of creosote is the one possible benefit of joint venting of a wood heater with a fossil fuel appliance.

Using separate flues is always safer, and I recommend it. But for the people who will or do make multiple use of a single flue, I offer the following advice:

1. If you have an open-type wood burner such as a fireplace or fireplace stove sharing a flue with another appliance, close both the flue damper and the doors or screens when the unit is not in use.

2. Be sure the flue has adequate capacity to handle all the connected appliances (see Table 2–2), particularly if one of the appliances burns oil or gas.

3. *Keep the flue clean.* This will prevent a decrease in chimney capacity due to creosote plugging, and will prevent chimney fires, making chimney fire suppression by air starvation unnecessary. The space inside the chimney below the lower breaching should also be cleaned out.

4. The two chimney connectors (stovepipes) should enter the common flue at different elevations by a few feet if possible. This will avoid one smoke stream pushing the other back.

Although there are conflicting opinions, most experts recommend that the wood appliance stovepipe enter the chimney *below* the oil or gas

Figure 2–29. Two ways to add a flue to (next to) an existing chimney, one safe, the other . . . well, you can see it has problems.

33

appliance pipe. There are two reasons for this. One is to keep the wood heater's potentially hotter stovepipe farther away from the ceiling. The other is that if creosote and ashes accumulate in the bottom of the chimney, the wood-burning appliance stovepipe would be blocked first. As explained earlier, blockage in the venting system of a wood-burning appliance is mostly an annoyance, not a safety hazard. Blockage of a gas or oil appliance is a serious hazard.

An alternative method of connecting two appliances to a single flue is to combine the two smoke streams with a stovepipe T or Y fitting *before* they reach the chimney. This requires large fittings and stovepipe since the diameter must be adequate to serve both appliances starting at the point where they join. This method is not recommended because of the subtle difficulties of doing it safely.

CHIMNEY CONNECTOR

Stovepipe is most commonly used as the "chimney connector" to connect wood burners to their chimneys. Not all installations have a chimney connector; factory-built fireplaces are often connected directly to their factory-built metal chimneys. Many fires result from unsafe chimney connectors.

Gauge

The standard material for chimney connectors is stovepipe—single-wall sheet-steel pipe. The minimum thickness steel, according to the National Fire Protection Association, is listed in Table 2–3.

It is not always easy to obtain the thickness or gauge pipe specified in that table.

Three reasons for using heavy gauge material are:

1. The higher mechanical strength and rigidity minimizes the chances of the pipe sagging, distorting, or moving. This is particularly important during chimney fires which can be quite violent *physically*.

2. The thicker gauge pipe will take longer to corrode through. Creosote can be quite corrosive.

3. The thicker walls will take longer to burn through. (High temperatures cause slow but inevitable oxidation or burnout of steel.) Thin gauge

pipe is *not* more likely to melt during a chimney fire; iron melts at about 2700° F., a temperature considerably higher than a bare exposed stovepipe could ever reach in almost any conceivable chimney fire.

The heavier gauge pipe recommended by NFPA is best, but if unobtainable, conscientious maintenance can compensate. With proper supports and sheet-metal screws the installation will have adequate mechanical strength. Intense chimney fires can be avoided by keeping the chimney and stovepipe clean. Frequent inspection, and replacement at the first sign of weakness, will prevent rusting and burnouts from becoming a safety problem. Comparing Canadian and U.S. standards in Table 2–2 makes it clear that gauge is not critical; respected opinions vary as to how thick is thick enough. (Connector pipe a full millimeter thick is not uncommon in European countries.)

To assemble ordinary stovepipe, two edges are joined to form a seam running the length of the pipe. Cutting to length is usually necessary during installations. This is relatively easy to do with shears before the seam is formed.

There is increasingly available higher quality stovepipe which has stronger, permanent seams through welding, metal bending, or riveting. This pipe is usually also appropriately heavy gauge (see Appendix 5 for sources). With the right kind of shears, or a hack saw, it also can be cut. Some manufacturers supply slip joints so that cutting is unnecessary when fitting pipe between a stove and its chimney.

Stainless steel stovepipe is difficult and expensive to obtain. It is substantially more resistant to corrosion and burnout, but unless painted will not emit as much heat as ordinary black or blue steel pipe.

All-fuel factory-built metal chimney sections can be used as chimney connectors *if* installed with no less than the clearances recommended by the manufacturer and with secure mechanical support. Some types such as the triple-wall air-cooled chimneys are designed to be installed vertically for convective cooling and should not be used for more than a few feet of horizontal run. Disadvantages of factory-built chimneys as connectors are the lack of heat output from them, the higher expense, and the esthetics, with the diameters 3 to 6 inches larger than the flue on the inside. However, if there is a need to pass a connector through a combustible wall, factory-built chimney sections can be very useful. Finally, Type L factory-built chimney sections are suitable as connectors.

Table 2-3. Connector Thicknesses

Minimum Suggested Thicknesses for Stovepipe Connectors
Used with Wood Heating Equipment[5]

| DIAMETER OF PIPE | NFPA (USA) RECOMMENDATIONS[1] | | CANADIAN RECOMMENDATIONS[3] | |
| | | | Galvanized Steel | Ungalvanized Steel |
(Inches)	Gauge No.[2]	Thickness (Inches)	Thickness (Inches)	Thickness (Inches)
Up to 5	26	0.019[4]	0.016	0.016
6	24	0.024	0.019	0.021
7	24	0.024	0.019	0.021
8	24	0.024	0.019	0.021
9	24	0.024	0.024	0.027
10	22	0.030	0.024	0.027

1. NFPA 211–1977, *Chimneys, Fireplaces and Vents 1977,* (Boston, Mass.: National Fire Protection Association, 1977).

2. These gauges are for *galvanized* pipe. NFPA does not give suggested gauges for ungalvanized pipe. It is reasonable to interpret the NFPA gauges as applying to either kind of pipe.

3. *Canadian Heating, Ventilating and Air-Conditioning Code 1977* (Ottawa, Canada: National Research Council of Canada, 1977).

4. In NFPA 211–1977, this particular thickness and the one given for 16-gauge material (0.058) are the only two that are inconsistent with Table A6-3 in Appendix 6. Why two differ and the other three agree, I do not know.

5. Typical connector pipe in Denmark is 1mm thick (about 0.039 in.).

Type L chimneys are designed principally for oil-burning appliances and should *not* be used as the *chimney* for a solid-fuel heater. But they are safe (and legal according to most codes) as chimney connectors for solid-fuel heaters and have the advantage over stovepipe of requiring less clearance to combustibles, 9 inches instead of 18.

Finishes

Four finishes on steel pipe are commonly available: blue steel oxide, black or other color paint, galvanization, and nickel "chrome" plate.

The finish has more effect on heating efficiency than on safety. Nickel-plated pipe in new shiny condition radiates very little heat. In a typical wood stove installation, with 6 feet of exposed stovepipe, one can expect up to 20 percent less heat output from the whole system when this type of pipe is used, compared to blue steel or paint finishes. New galvanized pipe is about half way between nickel-plated and black pipe in terms of radiating ability. Blue steel and pipe painted any non-metallic color are the best radiators.

Galvanized pipe should not be used in the living spaces of a house. The galvanization is mostly zinc, and zinc melts at about 750° F., a temperature often reached in a stovepipe. At these temperatures, toxic zinc vapor may be given off.

Stovepipe finishes have little direct effect on corrosion and burnout since these processes occur from the inside out and none of the finishes with the possible exception of nickel stands up very long on the inside of the pipe.

Although nickel-plated pipe radiates little heat, it should not be installed closer to a combustible surface than other pipes because installation will be safe only as long as the nickel-plate finish stays clean and shiny, which cannot be guaranteed. The finish discolors with use and age, resulting in more radiated heat, and if ever painted any non-metallic color, its non-radiating property is largely lost.

Visibility and Accessibility

Safety authorities and building codes presume that chimney connectors will rust or burn out and hence need occasional replacement. Even a conscientious homeowner may forget to inspect stovepipe if it is located in attics, under stairs, in closets, and even inside "legal" wall thimbles. Many fires have been caused by rusted or burned-out pipe the homeowner was not aware of because of its "hidden" location. Thus all stovepipe should always be fully visible.

Length

NFPA and most codes require chimney connectors to be as short as reasonably possible; thus the wood-burning appliance and its chimney must be as near to each other as possible. I consider lengths up to about 8 feet as normal for chimney connectors. Some authorities also recommend that the connector's horizontal run be no more than 75 percent of the chimney's height above the breaching.

The advantages of short chimney connectors are not all directly related to safety. Connectors are generally the weakest link, mechanically, of the system, and the longer they are the more susceptible they are to failure. Short stovepipe connectors also minimize creosote buildup and maximize draft by not letting the flue gas cool too much before entering the chimney.

The major advantage of long stovepipe runs is increased heat output. Stovepipe installations of 50 feet used to be common in churches and schoolhouses. The higher heating or energy efficiency is achieved by extracting heat, resulting in more creosote accumulation and less draft. Gutters and buckets were sometimes used to keep the liquid creosote off the floors, and weekly disassembly and cleaning out of the solid creosote was necessary. The chimney itself had to generate reasonably good draft on only warm flue gases,

and thus had to be an interior chimney, and neither too small nor too large in diameter. For closed stoves connected to good interior chimneys, horizontal runs of stovepipe connector *may* exceed 75 percent of the chimney height without the draft becoming inadequate.

Thus, as in so many other areas of wood heating, connector length involves compromises. The safest is definitely the shortest except in those rare cases where the combustion gases are so hot they *need* to be cooled to be safe for the chimney. Longer connectors can substantially increase the energy efficiency of the system but require considerably more care in the installation and more maintenance to be reasonably safe.

Mechanical Security

Stovepipe, regardless of its gauge, must be installed securely. The reason may seem obvious, but the amount of strength needed is more than one might expect. Chimney and stovepipe fires not only are very hot events, they also are physically violent, often resulting in substantial vibration of the stovepipe. Joints that may seem

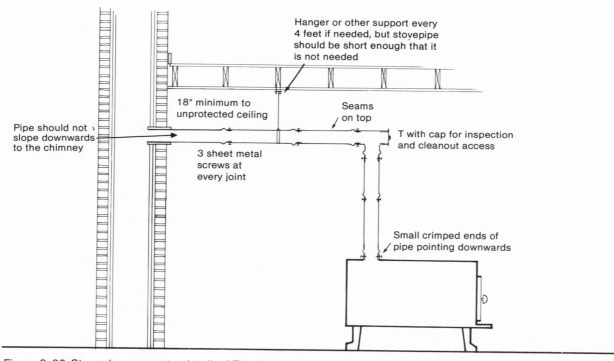

Figure 2–30. Stovepipe connector details. A T is shown in place of the usual 90° elbow; it permits easier inspection and cleaning of the stovepipe. T's tend not to be creosote-tight. Thus use at the bottom of a vertical section of connector may be inappropriate, depending on how the stove is operated. Most T's are not very streamlined; thus they may cause smoke spillage when used with open fireplace stoves.

adequate when just pressed tightly together without screws, are likely to vibrate apart during chimney fires. This should be avoided.

Stovepipe installations with a total length of about 8 feet or more should have extra mechanical support such as hangers or brackets, spaced about every 4 feet.

In *all* installations, *every* joint should be secured with at least three sheet metal screws or equivalent fasteners. The connection to the wood burner's collar must also be more than just a snug slip fit. Holes should be drilled through the collar to accommodate screws or bolts.

Connection to Chimneys

Manufacturers should plan for a safe and secure connection of stovepipe to their factory-built metal chimneys. A mechanically secure connection that also will not leak creosote is illustrated in Figure 2–31. If the manufacturer has not provided for an easy and secure connection, either buy a different brand of chimney or do the best you can—use screws, straps, bailing wire, or whatever it takes.

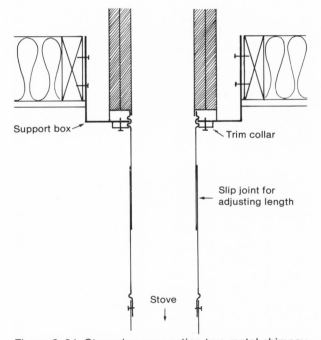

Figure 2–31. Stovepipe connection to a metal chimney. The details of the connection are different with different manufacturers. Illustrated here are some features in the Dura-Vent system. There are two independent fastenings—the upper lip on the top piece of stovepipe rests on the chimney support box; the second lip is supported by the trim collar, which is screwed into the support box. This system also keeps dripping creosote inside the system.

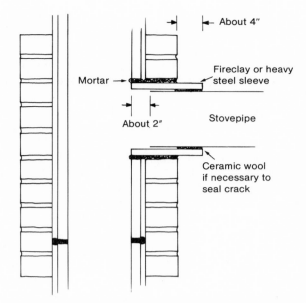

Figure 2–32. Stovepipe connection to a masonry chimney. The sleeve is mortared or cemented into place, but the stovepipe should be removable for inspection and cleaning of both the stovepipe and the chimney.

Safe connections of stovepipe connectors to masonry chimneys usually require more effort than connections to metal chimneys. If it is not already available, a hole must be cut through the chimney, with minimum damage to the chimney liner. Use of masonry drills can be helpful. The hole should be sized to accept a sleeve, liner, or thimble of fire clay, of about 3/16 inch steel, or of about 26 gauge stainless steel. The sleeve should extend through but not beyond the chimney's liner, and should extend a few inches into the room from the chimney's exterior surface (Figure 2–32). It should be permanently secured in the hole with either mortar or furnace cement.

The possibility of steel sleeves sliding in or out despite the mortar or cement can be minimized by selecting a sleeve with circumferential ridges. When using a thin gauge stainless steel liner, you can also flare out the inner end around the chimney liner with a small hammer. Care must be taken not to damage the liner.

The stovepipe should extend most of the way into the sleeve, but not protrude into the chimney flue. No additional mechanical fastening is necessary if the rest of the stovepipe installation is secure. The maximum possible horizontal movement of the stovepipe should not be more than 1 or 2 inches, not enough to come back out of the chimney or to protrude into the flue. If the sleeve is a little oversize for the stovepipe, the space between the two should be sealed, but not too permanently, as stovepipe always requires occasional replacement, and one needs to be able to pull out the pipe easily to inspect for creosote and soot in the chimney. Ceramic wool (see Appendix 5) stuffed into the crack is a good way to seal it. Asbestos rope works as well, but use of asbestos should be avoided when possible. Fiberglass insulation is not designed for such high temperature applications. It can smoke (it is not 100 percent glass), it can lose its resilience, and it can melt in the event of a hot chimney fire.

If the stovepipe cannot be inserted through a metal sleeve as described above, it can be fastened with screws to the protruding part of the liner. Sizing the liner so that the stovepipe *can* slip all the way through has the advantage of lowering the heat of the masonry and thus lessens the chance of its developing cracks due to thermal stress.

Elbows

NFPA, and therefore many codes and manufacturers' installation manuals, allows no more than two 90° elbows in the connector, and often specify that the elbows be smooth, with rounded corners, not sharp, angular turns. These requirements are related to performance more than to safety. All turns and corners in connectors and chimneys increase the resistance to flow for the flue gases, and sharp elbows can have twice the resistance of smooth elbows.

Does this added resistance matter? It usually depends on what type of wood burner is being used. For open burners such as fireplaces and Franklin stoves, extra resistance can often be critical. The system may not draw well enough to keep smoke from spilling into the house from the opening. Consequently the units either cannot be used, or can be used only with their doors shut. The reason is the large required air flow through the opening to keep the smoke in, and the fact that fireplace and fireplace stove chimneys are not usually designed with much excess capacity.

Most chimneys have excess capacity when serving *closed* burners. The resistance to air flow at the stove's air inlet is usually dominant. Adding a little more resistance with an extra elbow usually is not significant. In addition, even if the added resistance to flow *is* significant, the result is not an unsafe installation, but one that may spill some smoke into the room during refueling, or one whose peak heating rate is less. With less air being pulled into the stove the fire is not as vigorous, and so less heat is put out into the room.

Upward Slope

NFPA requires horizontal runs of stovepipe or any chimney connector to have a pitch of at least a ¼-inch per foot of pipe, sloping up from the woodburner to the chimney.

This upward slope is intended to assure good draft, and is based on the notion that hot gases will not flow downhill. In practice an upwards slope is rarely important. Hot gases will readily flow horizontally or even downhill as long as there are sufficient vertical portions in the stovepipe and chimney to provide the suction or push required.

Significantly downward sloping pipe should be avoided. Draft will be slightly decreased. Creosote dripping will be harder to control. Ashes and creosote may accumulate at a low point and block flow. (Actually, an upwards slope of ¼-inch per foot is so slight that similar blockage can occur just as easily whether or not the slope is present, particularly at a horizontal-to-vertical elbow.)

Creosote Drip Potential

Creosote has an unpleasant, acrid odor, it stains, and is flammable after it dries. Thus it is both safer and more pleasant to keep creosote inside the wood heating system. The most serious problems with creosote dripping usually occur with the chimney connector.

Creosote can always be minimized by careful operation of the wood burner. One can also use pipe and fittings designed to contain creosote. Some practical suggestions for creosote containment follow:

1. Vertical and upward-sloping stovepipe joints should have the smaller-diameter, crimped end sticking down into the larger end (Figure 2–30) to keep the liquid inside the system as it flows down. Contrary to intuition, this will very rarely result in smoke leaking out the joints.

2. The stovepipe should fit *inside* the collar. This may require a stovepipe adapter or reducer since many wood heaters are made to receive standard size stovepipe outside the collar. However a number of thoughtful manufacturers produce stoves designed for the stovepipe to fit inside the collar.

Since fitting stovepipe *inside* the collar on wood heaters is often difficult, placing it outside and carefully applying a good high temperature cement or sealant is a possible but usually less

38

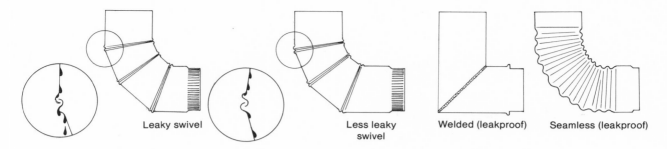

Leaky swivel   Less leaky   Welded (leakproof)   Seamless (leakproof)
                    swivel

Figure 2–33. The creosote tightness of stovepipe elbows. The sharp corner of the welded elbow offers significant resistance to flue gas flow, so it should not be used in fireplace stove installations.

effective alternative. Some furnace cements shrink or crack and thus may not form a permanent seal. Two reportedly effective high temperature sealants are Silver Seal and Hercules ® Hi-Heat Furnace Cement (see Appendix 5 for sources). Whatever material is used, joints are often self-sealing in time due to creosote, ash, and corrosion products filling any voids.

In many cases the pipe-to-stove seal is not critical since temperatures are high enough to evaporate or dry any creosote which would be there.

3. Similarly the connection to the chimney should be drip-proof so as to contain the creosote *inside* the pipe. This is most often necessary when connecting vertical stovepipe into the bottoms of factory-built chimneys. Increasingly chimney and stovepipe manufacturers are making available special drip-proof fittings (Figure 2–31). When horizontal stovepipe connects to any kind of chimney, most liquid creosote formed in the chimney will not enter the chimney connector but will drip down beyond the connection towards the chimney cleanout area.

4. If stovepipe with leaky seams is used, the seams should be on the top side of the horizontal pipe. Seamless or continuously welded or other tight pipe is increasingly available and good to use, not only for its tightness, but because it tends to be heavier gauge.

5. Elbows are frequently sources of creosote leaks. One common swivel type of elbow has three circumferential seams joining four riveted bands of sheet steel (Figure 2–33). The seams are not liquid-tight because they are designed to allow rotation of the bands so that the elbow's angle is adjustable from 90° down to 0° (no bend). Three possible approaches to the resulting creosote leakage problem are: (a) Since the seams in swivel elbows are fairly tight, one can just wait until a little creosote, ash, or corrosion seals them up, (b) There are two types of swivel elbows. One can use swivel elbows where the downward end of each band fits inside the next lower band. (c) Use seamless elbows such as welded miter elbows (non-adjustable) or accordion elbows (adjustable).

6. Sealing the joints between pipe sections can be useful, particularly in horizontal runs. To be effective the sealant should be applied on the joint surfaces before assembly, not to the crack after assembly. Use of sheet metal screws at the joints will help the cement to form a permanent seal. Without them, the inevitable movement of the pipe will crack and loosen ordinary furnace cement. Many cements crack and loosen anyway.

Wall Pass-throughs

Many fires start in walls through which stovepipe passes. The best *simple* advice is "Don't do it." Locate the stove in the same room as its chimney. However if the venting system must pass through a wall, observe NFPA recommendations and applicable code requirements.

Most codes prohibit passing a chimney connector through a *ceiling* or *floor* for two reasons. Bare stovepipe requires 18 inches of clearance to combustible materials, and stovepipe at floor level (above the ceiling) is more susceptible to clearances around it being changed as furniture and rugs are moved about. Also fires spread more quickly and more lethally through holes in ceilings than through holes in walls.

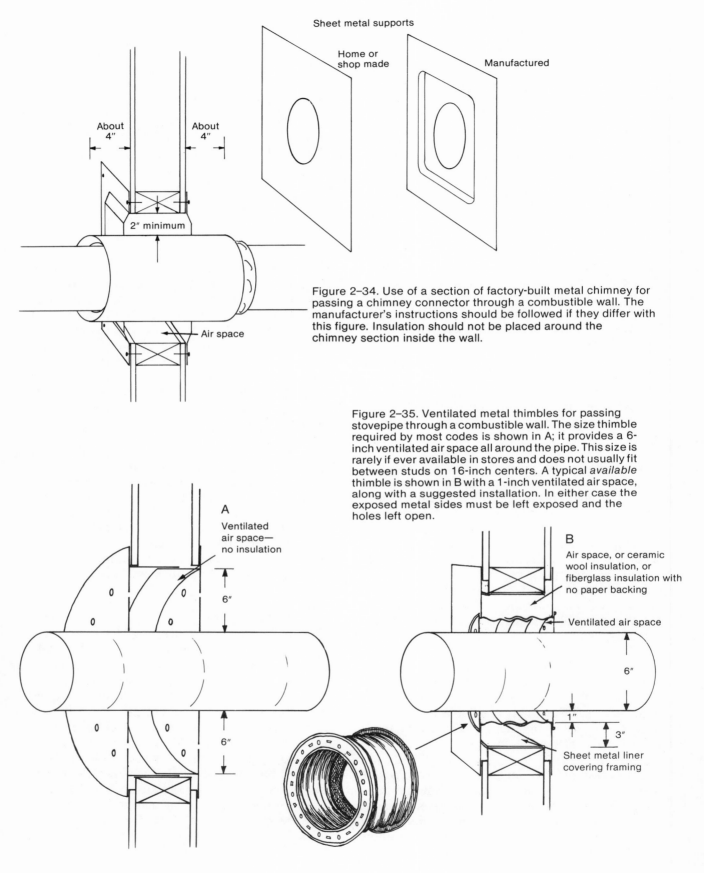

Sheet metal supports

Home or shop made

Manufactured

About 4"

About 4"

2" minimum

Air space

Figure 2–34. Use of a section of factory-built metal chimney for passing a chimney connector through a combustible wall. The manufacturer's instructions should be followed if they differ with this figure. Insulation should not be placed around the chimney section inside the wall.

Figure 2–35. Ventilated metal thimbles for passing stovepipe through a combustible wall. The size thimble required by most codes is shown in A; it provides a 6-inch ventilated air space all around the pipe. This size is rarely if ever available in stores and does not usually fit between studs on 16-inch centers. A typical *available* thimble is shown in B with a 1-inch ventilated air space, along with a suggested installation. In either case the exposed metal sides must be left exposed and the holes left open.

A
Ventilated air space— no insulation

6"

6"

B
Air space, or ceramic wool insulation, or fiberglass insulation with no paper backing

Ventilated air space

6"

1"

3"

Sheet metal liner covering framing

40

If a chimney connector must pass through a wall containing any combustible material, structural, finish or insulating, the following are safe ways to do it, according to NFPA:

1. Use factory-built, listed (such as by Underwriters Laboratories) chimney sections and install them in accordance with the manufacturer's instructions (Figure 2–34). *No common factory-built metal chimneys are well enough insulated to be safe when touching combustible material.* Typically, at least two inches of clearance are required. Some factory-built chimneys are not intended to be used horizontally; these do not satisfy NFPA's requirements. If brands intended for horizontal use are not easily available, I suggest using insulated chimney, since this type is less dependent on air for cooling.

2. Use a metal ventilated thimble with at least a 6-inch larger radius than the connector stovepipe (Figure 2–35). For a 6-inch diameter stovepipe, the thimble must have an 18-inch diameter.

Such large ventilated metal thimbles are not readily available. Most thimbles on the market have only about 2 inches between the pipe and the thimble wall instead of the 6 inches recommended by NFPA. Thus, in terms of satisfying many building codes, one must either have a thimble fabricated, or not use this method for passing stovepipe through walls.

Whether the available thimbles with 1- to 2-inch ventilated space are safe is another question. If they are not safe, either they should not be on the market or they should be clearly marked as not intended for use in wood stove installations. If they are safe, the codes prohibiting them should be changed. But lacking any hard evidence, the conservative choice is appropriate where safety is concerned. Thus I would not trust these thimbles by themselves, but would recommend installing them with an additional margin of safety. For instance, to pass a 6-inch stovepipe through a wall with studs on 16-inch centers, I would cut a 14½-inch square or round hole between studs (cut the hole flush with the two facing stud sides), cover the exposed wood with sheet steel or aluminum flashing, use sheet steel to cover the hole up to the thimble, and put either loosely packed ceramic wool or vermiculite between the thimble and the studs.

There are two disadvantages to any ventilated thimble pass-through. If passing through an exterior wall, the ventilation results in a heat loss just as do leaks or cracks anywhere in the house. The ventilation is necessary for use of the thimble to be safe. Another possible disadvantage is the lack of visibility of the stovepipe inside the thimble. Should that part of the pipe be rusting or burning out there would be no visible evidence. Because of these problems, particularly the heat loss due to the ventilation, I prefer other methods when passing stovepipe through an exterior wall.

3. Stovepipe may pass through a combustible wall if a large area around the pass-through is made into a noncombustible masonry wall. Specifically, the stovepipe must be surrounded on all sides by at least an 8-inch thickness of mortared bricks with a metal or fireclay sleeve embedded in the center (Figure 2–36). In this case, even if the stovepipe should leak creosote or become weak inside the wall, the wall itself is not directly threatened. Note that a 6-inch diameter pipe requires a $8 + 6 + 8 = 22$-inch square or diameter hole to be cut out and rebuilt in masonry. Since some kinds of natural stone have considerably higher thermal conductivity than typical bricks, stone should not be assumed to be equivalent to brick.

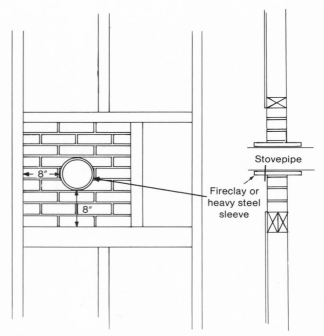

Figure 2–36. A masonry-patch, wall pass-through for stovepipe. The sleeve should be permanently fixed in place with mortar or furnace cement. The sleeve should also extend 4 inches beyond the masonry on both sides. The masonry should be left exposed.

4. Finally, if all combustible material is removed from around the stovepipe to the clearance required for stovepipe out in the open, the pass-through is considered safe (Figure 2–37). This distance is usually taken to be 18 inches. Thus a hole 18 + 6 + 18 = 42 inches wide is required for 6-inch stovepipe. The hole can be covered with any noncombustible and high-temperature-tolerant material such as sheet metal, asbestos cement board, or asbestos millboard. The space inside need not be ventilated and may be insulated, but neither paper nor foil backings should be on any fiberglass insulation used in this way. Vermiculite or ceramic wood is a better insulator to use next to a hot stovepipe. This is generally the least practical method of passing stovepipe through a wall. The hole size specified by NFPA is so large it may weaken the wall unless properly reinforced. It is also unlikely to enhance the esthetics of a room.

These same basic principles apply when connecting stovepipe to a masonry chimney that is covered by a wall or faced with combustible materials. For instance some chimneys may have wood lathing and plaster on or very near them or they may be boxed with paneling or ordinary stud walls. One of these four described methods should be used to protect the combustible material where the connector passes through to the chimney.

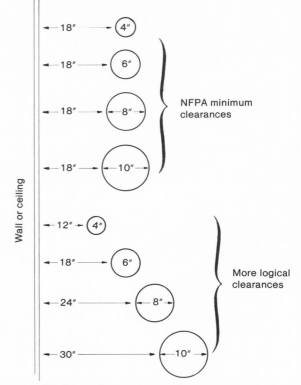

Figure 2–38. Stovepipe minimum clearances to unprotected combustible walls or ceilings. Both sets of clearances are reasonably safe. Almost all building codes specify 18 inches of clearance regardless of stovepipe diameter.

Clearance

NFPA and most codes require that stovepipe chimney connectors be at least 18 inches away from any parallel combustible wall or ceiling. A wood frame wall or ceiling covered with plaster or gypsum board is classified as combustible. An alternative standard often discussed but rarely incorporated into codes is a clearance of three times the stovepipe's diameter (Figure 2–38).

If all stovepipe sizes had the same surface temperature, larger pipe would require a larger clearance because it would radiate more heat. However it can be argued that in many cases installations with larger pipe diameters *tend* to have cooler pipe temperatures due to more excess air or better heat transfer. But I suspect that *peak* stovepipe temperatures do not correlate with pipe size. Thus a size-dependent clearance seems to be logical. For 6-inch pipe, the two standards above give the same clearance—18 inches.

Note that reducing the stovepipe size down from the flue collar size in order to allow closer

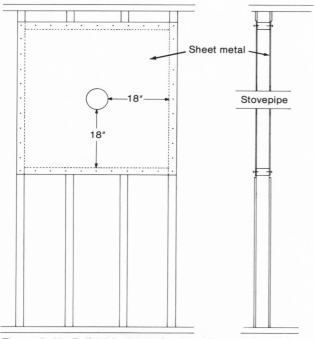

Figure 2–37. Full 18-inch, cut-back wall pass-through for stovepipe.

Figure 2–39. Consequences of inadequate protection of wall around a penetrating stovepipe and factory-built chimney. Much of the charred wood was cut away by the firemen. The junction between the stovepipe and the insulated metal chimney was inside the wall, an unusual but not necessarily dangerous situation. The serious problem was lack of clearance around both the stovepipe and the metal chimney. Direct contact between the chimney and the wood siding is visible.

distances to a wall or ceiling is not generally a viable option. It is not only illegal under most codes but also the appliance may not heat the house very well due to inadequate draft.

Reduced clearances are safe if the wall or ceiling is adequately protected, as described in a following section in conjunction with reduced stove clearances.

Stovepipe Damper

Manually operated stovepipe dampers (Figure 2–40) are a traditional part of wood heater installations. In the case of non-airtight wood heaters they are useful for reducing the air flow into the combustion chamber, thus obtaining longer duration and lower power output burns.

In the case of *airtight* stoves such dampers are unnecessary for this objective since the air supply can be adequately limited by the air inlet controls on the stove itself. In some cases, stovepipe dampers are useful additions to airtight equip-

ment. Many users of airtight stoves believe that closing a stovepipe damper results in more energy-efficient operation and so more total heat for the same amount of wood burned. Tests in the author's wood heating laboratory have verified this effect for one particular stove, the Energy Harvester, in some limited testing. Both combustion efficiency and heat transfer efficiency were improved. Closing the stovepipe damper usually requires increasing the opening of the air inlet damper in order to maintain the same power output.

For safety, stovepipe dampers should not be capable of completely blocking the flow of smoke. Standard dampers have holes through them. They are also made not to fit tightly inside the stovepipe. Nonetheless closing them may still offer some assistance in limiting the intensity of chimney fires, particularly with non-airtight stoves.

It is not critical exactly where along the stovepipe the damper is installed.

Barometric Draft Regulators

Some codes and NFPA require draft regulators in the chimney connector of wood or coal furnaces and boilers, and manufacturers recommend their use with some stoves. There is some doubt among wood heat experts as to whether use of barometric draft regulators is wise.

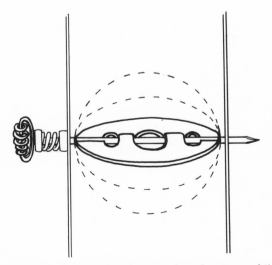

Figure 2–40. A stovepipe damper. An advantage of the inherent leakiness of this damper is that smoke is less likely to back up into the house when the damper is closed, whether by malfunction, by mistake or by misguided intention. A disadvantage is less control of both normal stove fires and of chimney fires, particularly when the appliance itself is not very airtight.

Draft regulators limit the draft—the suction pulling air into the appliance. A pivoted, counterbalanced flap is pulled open by the draft when the draft reaches a critical amount (Figure 2–41). This permits air to enter the chimney, thus preventing the draft in the appliance from rising any higher.

Use of draft regulators with oil-fired equipment is common. One objective is to keep the oil flame from being blown out by excessive draft. This is more an annoyance than a safety problem since most oil burners have primary safety controls which stop the fuel flow into the burner whenever the flame goes out, and modern flame-retention burner design makes draft regulation almost unnecessary for this objective.

A second objective is to limit the amount of air passing through the system in order to maintain high heat transfer efficiency. Another desirable consequence is a decrease in peak chimney temperatures due to the added air entering through the regulator.

High draft does not blow out the flames but fans them in wood- and coal-burning equipment. An especially hot fire can result. Control of the fire may be difficult, particularly if the wood burner is not airtight. Energy efficiencies may be adversely affected, and chimney fires may be ignited. A barometric draft regulator can alleviate these problems by limiting the flow of combustion air into the appliance and by adding cooling air to the hot flue gases.

However the cooling air may result in more creosote accumulation, and if a chimney fire *does* start, it will be more intense because of the draft regulator. Large flows of air into the regulator will fan the fire in the chimney. Thus the net effect of barometric draft regulators is unclear—they help in some aspects, and hurt in others. Note, though, that if a chimney is clean so that chimney fires are

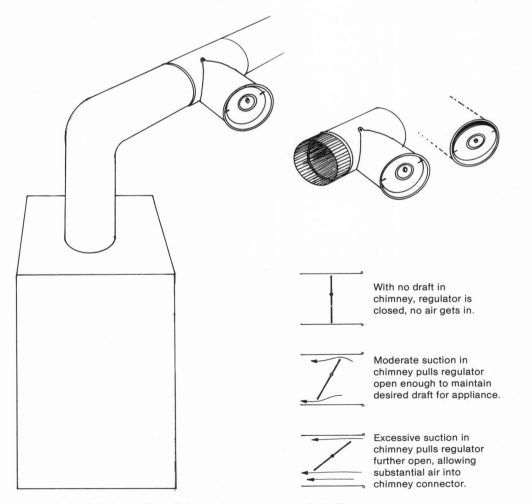

With no draft in chimney, regulator is closed, no air gets in.

Moderate suction in chimney pulls regulator open enough to maintain desired draft for appliance.

Excessive suction in chimney pulls regulator further open, allowing substantial air into chimney connector.

Figure 2–41. A barometric draft regulator and its intended effect.

44

impossible, a draft regulator does no harm, and does help limit the intensity of the fire in the appliance, and it helps prevent the flue gases in the chimney from getting too hot. Barometric draft regulators are recommended by NFPA for hand-fired thermostatically controlled solid fuel furnaces.

Other Stovepipe Draft-Regulating Systems

One technique used to regulate draft with non-airtight stoves is to install a T in the stovepipe with a manual damper to regulate the amount of room air let into the stovepipe (Figure 2–42). Leaky stoves may burn up their fuel more quickly than desired. Letting air into the stovepipe results in less suction pulling air into the stove through the stove's leaks and hence slower burning of the fuel. A barometric draft regulator set for very low draft would accomplish about the same objective as this manual draft regulator set open.

Safety problems are created with this installation. Since draft (suction) is not regulated automatically, as it is with a barometric draft regulator, draft control is more irregular. Because it is a hole in the venting system, sparks, creosote and smoke may get out through it into the house. Barometric draft regulators are less susceptible because of their horizontal orientation and because they close and seal tightly when drafts are less than the set point, such as when a wind gust might force smoke out of the systems in Figure 2–42. In case of a chimney fire, any of these draft regulators fan the flames.

I strongly oppose using non-automatic chimney draft regulators. Safer options are to accept the fact that long duration burns are not obtainable from most leaky stoves, or to try to seal the leaks, or to buy a tighter stove.

Stovepipe T's

Easy access for inspection and cleaning *inside* a stovepipe chimney connector is important, for the easier it is to do, the more likely it is that it will be done. Thus in some installations it may be useful to install a T fitting in the connector (Figure 2–30). However, T's should not be used in most fireplace installations because their corners may offer too much resistance to smoke flow. They also should not be used at lower points in systems where wet creosote is common, as they are rarely watertight; thus creosote may drip through them onto the floor.

Figure 2–42. Relatively unsafe draft regulating schemes, particularly system B.

THE WOOD HEATER ITSELF

The two major areas of concern when installing the wood heater are clearances or protection to prevent walls and floors from getting dangerously hot due to radiated heat from the appliance, and assuring that the inevitable sparks or glowing coals that occasionally fall out of the appliance do not ignite the floor.

If a wood-heating appliance is a listed appliance, having been tested for safety by a nationally recognized laboratory, all required clearances and floor protection should be spelled out in the manufacturer's instructions. The appliance model has been proven in actual tests not to cause dangerously high temperatures of nearby combustible materials, when installed in accordance with the manufacturer's instructions, and when operated normally. *These "listed" clearances then supersede the following guidelines.*

Factory-built fireplaces are virtually all "listed" and thus come with their own installation instructions. Wood-fired furnaces and boilers are discussed in Chapter 5. The following discussion focuses on stoves, fireplace stoves, and cooking stoves.

Clearances

Table 2–4 gives the minimum clearances that NFPA recommends from wood heaters to combustible materials. Virtually all common wall types except 100 percent masonry are considered combustible.[3] A gypsum board or plaster finish on a wood stud wall is considered a "combustible structure" because the gypsum board and plaster conduct heat so well that in steady state its side in contact with the studs will be almost as hot as its exposed side.

Essentially, 36 inches is the required clearance to walls for radiant stoves, and 12 inches for circulating or jacketed type stoves.

A stove should not be installed where it could overheat a door left open. If the required clearance

---
3. The National Fire Protection Association defines combustible material as follows: "Material made of or surfaced with wood, compressed paper, plant fibers, plastics, or other material that will ignite and burn whether flameproofed or not, or whether plastered or unplastered." NFPA 97M, "Glossary of Terms Relating to Heat Producing Appliances."

to combustibles is 36 inches, stoves should be installed at least 36 inches from the arc of any wooden door.

The basic clearances of 36 and 12 inches are intended to be adequate for all stoves and will be more than necessary for many stoves, particularly small ones and stoves whose designs result in less-hot surface temperatures. This is discussed in detail in Appendix 2. For instance, if 36 inches is a safe clearance for the side of a Riteway 37, then 20 inches is an equally safe clearance for the back of a Jotul 602, assuming equal surface temperatures of the stoves. The area of the Riteway's side is about 5.5 square feet, whereas the back of the Jøtul 602 has an area of only about 1 square foot. Looked at another way, if the Riteway is safe at 36 inches for an average surface temperature of up to 700° F., the Jøtul 602 is equally safe at 36 inches for an average surface temperature of up to about 1200° F., which is unachievably high with wood fuel.

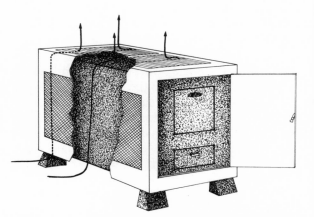

Figure 2–43. A circulating-type wood or coal heater. The double-wall construction with circulating air in between results in cooler exterior temperatures, permitting installation closer to combustible walls.

Figure 2–44 can serve as an approximate guide for safe clearances of stoves from unprotected combustible walls, as a function of the size of the stove's side that is facing the wall. To achieve equal degrees of safety, small stoves *can* be placed closer to walls. Figure 2–44 assumes the stove-pipe chimney connector does not attach to the stove on the side facing the wall. If it does, then chances are the minimum safe clearance of the stove from the wall will be determined by minimum safe *stovepipe* clearances. Note that Figure 2–44 does not apply to stoves operated so hot they glow.

Table 2-4. NFPA Clearances

**NFPA-Recommended Clearances
for Unlisted Wood Heating Equipment, and Stovepipe[1]**

	From Sides	*From Back*	*From Front*	*Above Top*
Radiant stove	36 inches	36	36	36
Circulating stove	12	12	24[2]	36
Cooking stove	—	36[3]	—[4]	30
Firing side	36[3]			
Opposite side	18			
Stovepipe	18[5]			

---

1. "Heat-Producing Appliance Clearances," NFPA 89M-1976 (Boston, Mass.: National Fire Protection Association, 1976).

2. The "front" or loading-door side should probably be given more clearance than 24 inches just for easy access. The 24 inches listed is probably an estimate of the minimum space needed for convenience of operation. Since to be called a "circulating stove" the stove must be fully jacketed, including the door side (according to NFPA's definition), the stove is unlikely to be any hotter on the "front" side than on any other side, and thus would not need larger clearance for safety. However, floor protection should extend 18 inches out on this side.

3. If the cookstove has a "clay-lined firepot" (a firebrick or other refractory-lined combustion chamber), the back side clearance need only be 24 inches and the side clearance on the firing side need only be 24 inches.

4. NFPA presumes the front side will have plenty of safety clearance due to naturally provided access clearance. (One can reasonably surmise from the given side and back clearances that 36 and 24 inches would be the minimum front safety clearances for unlined and lined combustion chamber cookstoves.)

5. This clearance is to a *parallel* surface, such as between vertical pipe and a wall, or between horizontal pipe and a ceiling. For required protection when stovepipe passes through a wall (and hence is perpendicular to the wall), see text.

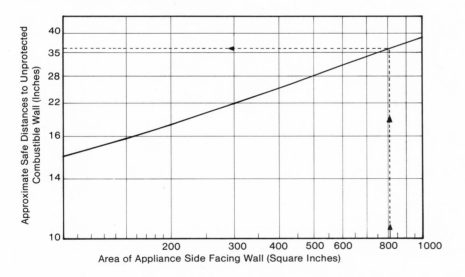

Figure 2–44. Approximate safe clearances of stoves from unprotected combustible walls, as a function of stove size. To use the graph, multiply the height and width in inches of the stove side facing the wall. Find this number along the bottom of the graph. Follow straight upwards to the diagonal line, and then left to the left edge of the graph. The number there is the safe distance of the stove from a wall. In the example, the side of the Riteway Model 2000 is 812 square inches, so the graph indicates 36 inches as safe clearance.

The graph is based partly on NFPA guidelines but mostly on theoretical calculations (Appendix 2). The theory was used to determine the slope of the line. The line was located to give a 36-inch clearance (NFPA's safe clearance for all radiant stoves) for the side of the Riteway Model 2000, one of the largest well-known stoves.

Stovepipe clearances must also be satisfied, and this may require the stove to be placed further away from a wall than the distance the graph indicates.

This graph does not apply to stoves operated red hot.

Any stove with an external baffle or shield spaced out at least half an inch from the stove and that allows air to circulate freely between it and the stove can typically be safely placed closer to a combustible wall than the same stove without the baffle. Internal structures in stoves can also affect outside surface temperatures and safe clearances, but predicting safe reduced clearances, without actual measurements, is more difficult; liners, some kinds of baffles, and some air inlet designs *may*, but do not necessarily, reduce stove surface temperatures.

Code writers have chosen not to take these variations into account, opting instead for brevity and simple enforceability. These are important objectives in codes, but the result is inflexibility with no safety rationale. Codes are often compromises between simplicity and technical rationality.

Reduced Clearances

NFPA guidelines and most codes permit reduced clearances to combustible walls and ceilings if adequate protection is added. A very common mistake is to presume that sheet metal, masonry, or asbestos board placed directly against a wall protects it (Figure 2–45). In fact, these materials installed in this manner afford very little protection. Although shiny sheet metal does offer some protection because of its infrared reflectivity, it should not be relied upon because aging or painting can too easily destroy its reflectivity.

These materials conduct heat very well—they will be almost as hot on their back sides as they are on their exposed sides. Thus the wall can get hot enough to smolder and burn (Figure 2–46).

A wall can be kept cool using sheet metal, thin masonry, and asbestos board, but only if they are mounted spaced out from the wall by an inch or more, to allow free circulation of room air behind the protective panel (Figure 2–47). *The circulating air keeps the wall cool by carrying away the heat from the space between the wall and the protecting panel.* The protector itself must of course be noncombustible because *it will* get hot.

How much closer a stove may be to a *protected* combustible wall is specified by NFPA and most codes. Table 2–5 lists all NFPA-permitted wall protections and corresponding clearances. My own recommendations differ somewhat from those of NFPA and are given in Table 2–6. Typically a ventilated wall protector permits a radiant stove to be placed 12 inches from the

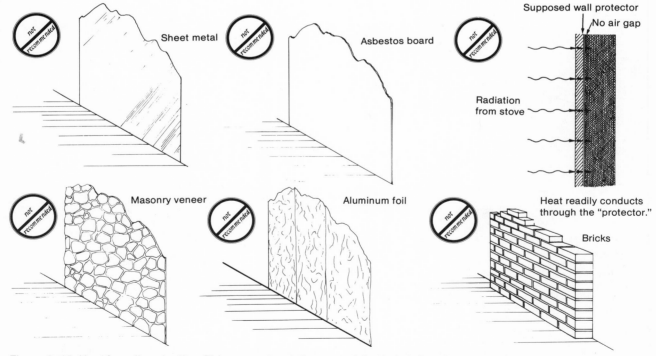

Figure 2–45. Unsafe wall protection. Thin or non-insulating materials placed directly against walls give little protection. With these installations, stoves should be placed no closer to the walls than if the materials were not there. A brick facing offers some protection, but without a ventilated airspace behind it, many safety experts consider the protection small.

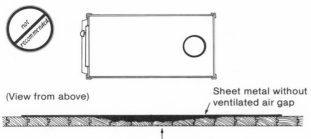

(View from above)

Sheet metal without
ventilated air gap

Charred wood wall

Figure 2–46. The consequences of inadequate wall protection. The sketch illustrates the problem with unventilated thin non-insulating protectors. The photographs illustrate a real case. (Photos courtesy of Leonard Fontaine) In photo A the wall "protector" was masonry veneer applied to a sheet of plywood, nailed to the paneled wall. The stove was less than 6 inches from the wall. B shows a small dark stain where smoke from the smoldering plywood came through a gap in the mastic-type adhesive. C shows the back of the plywood charred so severely that there are small holes through the sheet. The paneling behind the "protector" was charred completely away, revealing (in photo D) the fiberglass insulation in the wall.

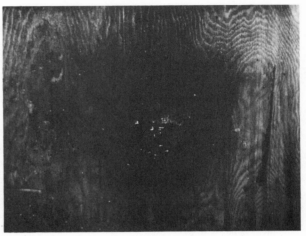

C

A

B

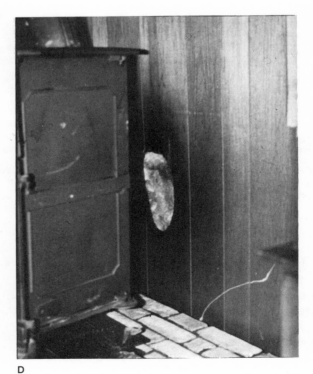

D

Table 2-5. NFPA Reduced Clearances

NFPA Reduced Clearances with Specified Forms of Protection According to NFPA[1,2,3,4]

TYPE OF PROTECTION

Applied to the combustible material and covering all surfaces within the distance specified as the required clearance with no protection. Thicknesses are minimal.	REQUIRED CLEARANCE WITH NO PROTECTION					
	36 Inches		18 Inches		12 Inches	9 Inches
	Above	*Sides & Rear*	*Sides & Rear*	*Chimney Connector*	*Sides & Rear*	*Chimney Connector*
(a) ¼-in. asbestos millboard spaced out 1 in.[3]	30	18	9	12	6	6
(b) 0.013-in. (28-gauge) sheet metal on ¼-in. asbestos millboard, no air gap	24	18	9	12	6	4
(c) 0.013-in. (28-gauge) sheet metal spaced out 1 in.[3]	18	12	6	9	4	4
(d) 0.013-in. (28-gauge) sheet metal on ⅛-in. asbestos millboard spaced out 1 in.[3]	18	12	6	9	4	4
(e) ¼-in. asbestos millboard on 1-in. mineral wool bats reinforced with wire mesh or equivalent, no air gap[5]	18	12	6	6	4	4
(f) 0.027-in. (22-gauge) sheet metal on 1-in. mineral wool bats reinforced with wire or equivalent, no air gaps[5]	18	12	3	3	2	2

1. Adapted from NFPA 89M-1976, "Heat-Producing Appliance Clearances" (Boston, Mass.: National Fire Protection Association, 1976).

2. All clearances should be measured from the outer surface of the appliance or chimney connector to the combustible material disregarding any intervening protection applied to the combustible material.

3. Spacers should be of noncombustible material. These methods a, c, and d require ventilation between the sheet material and the protected combustible material.

4. Asbestos millboard referred to above is a different material from asbestos cement board. It is not intended that asbestos cement board be used in complying with these requirements when asbestos millboard is specified.

5. The mineral wool bats should have a minimum density of 8 pounds per cubic foot and a minimum melting point of 1500° F.

original wall surface (Figure 2–48); however for stoves with side or rear exiting stovepipe, the distance from the wall must be greater and is determined by the minimum stovepipe clearance from protected walls (typically 9 inches, but see Tables 2–5 and 2–6).

Wall protectors must be large. NFPA requires all the wall area within 35 inches of a stove to be covered by the wall protector. This will typically require the protector to be 6 or 7 feet wide, and 5 to 6 feet tall (more if stovepipe protection is also required).

It is my opinion that this NFPA requirement is excessive (see Appendix 2). The following are the guidelines behind my recommended wall protector extents illustrated in Figure 2–48:

Table 2-6. Reduced Clearances (Author's Recommendations)[1]

	PRINCIPAL APPLICATION	Radiant Stoves		Stovepipe		Circulating Stoves	
	TYPE OF PROTECTION	RECOMMENDED CLEARANCE WITH NO PROTECTION					
		36 Inches		18 Inches		12 Inches	
	Dimensions are minimal.	As Wall Protector	As Ceiling Protector	As Wall Protector	As Ceiling Protector	As Wall Protector	As Ceiling Protector
	4-in. thick (nominal) brick or stone facing without a ventilated airspace	24	—	12	—	9	—
	A 1-in. ventilated airspace	12	18	6	9	4	6
	Two 1-in. ventilated airspaces	6	12	3	6	3	3
	Sheet metal on 1 in. of insulation mounted with a 1-in. ventilated air gap between the insulation and the wall or ceiling	6	9	3	4	3	3

Clearances are measured from the outer surfaces of the appliance or stovepipe to the *original* wall or ceiling surface.

In no case should the appliance or pipe *touch* the protector; there should at least be a 1-inch gap.

All materials including fasteners should be non-combustible, be adequately strong and retain their strength and insulating properties at high temperatures. Two recommended ways to obtain a one inch ventilated airspace are using sheet metal, and using bricks, as discussed in text.

A circulating stove is a stove *fully* jacketed in sheet metal providing a ventilated airspace all around the four sides and top of the stove. There will be perforations or louvers or registers in the upper sides or top of the jacket to let the hot air out.

1. These clearances and protectors are not necessarily approved by NFPA or by local building codes.

For no-protection clearances other than 36, 18 and 12 inches, linear interpolation may be used. This situation may arise if the size-effect of appliances is taken into consideration. For instance, if a stove has a 400-square-inch side facing a wall, then the safe clearance with no protection is 27 inches. This is halfway between the 18- and 36-inch no-protection-clearance columns in this table. Thus the appropriate reduced clearance with a ventilated airspace protector is halfway between 6 and 12 inches or 9 inches.

These clearances are not intended to apply to heaters operated so hot that their exteriors glow red hot continuously. Such operation is always damaging to the heater.

1. Wall protectors should start within a few inches of floor level.

2. Lateral and vertical extent should be at least 18 inches (9 inches for a circulating stove) beyond any part of the appliance's shadow (that is, 18 inches to the sides of and above the perpendicular projection of the stove onto the protected wall.)

Additional protection may be needed higher up the wall due to the stovepipe.

3. In the case of fireplaces or fireplace stoves oriented so that the wall to be protected can receive radiation directly from the fire, wall protection should extend 30 inches beyond the fireplace opening (i.e., 30 inches beyond the perpen-

51

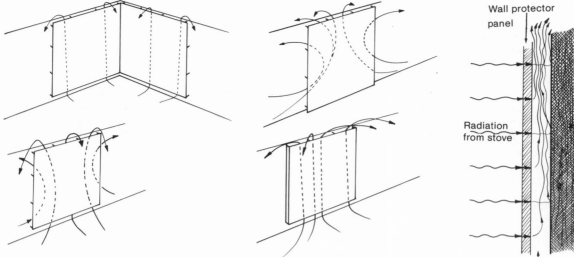

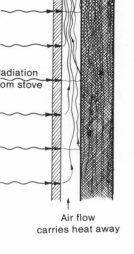

Wall protector panel

Radiation from stove

Air flow carries heat away

Figure 2–47. Ventilated wall protectors, the best kind of wall protection.

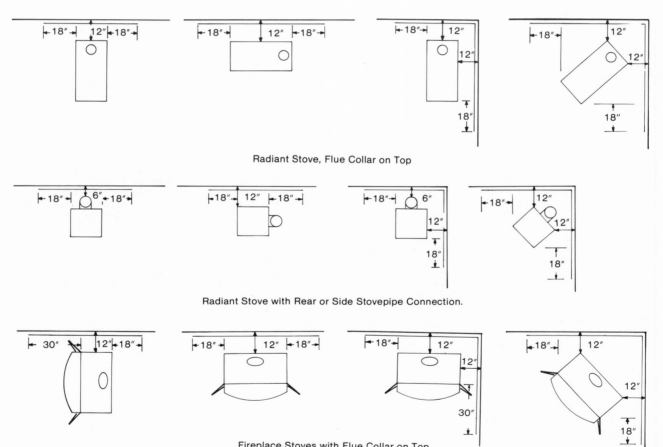

Radiant Stove, Flue Collar on Top

Radiant Stove with Rear or Side Stovepipe Connection.

Fireplace Stoves with Flue Collar on Top.

Figure 2-48. The author's (not the NFPA's) suggested minimum extents of ventilated-gap wall protectors for radiant stoves and for fireplace stoves. If the stovepipe connects to the back of the fireplace stove, the vertical stovepipe must be no closer than 6 inches to the wall. These protector extents do not satisfy most building codes.

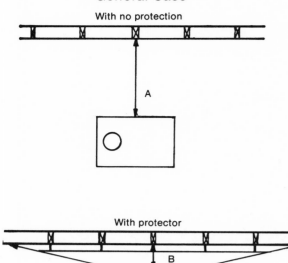

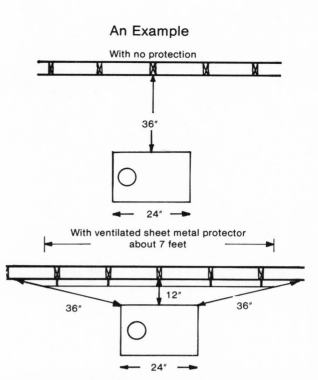

General Case

With no protection

A

With protector

C   B   C

C must equal A.

An Example

With no protection

36″

24″

With ventilated sheet metal protector
about 7 feet

12″

36″   36″

24″

Figure 2-49. NFPA requirements for the size of wall protectors. If A is the required minimum clearance to an unprotected combustible wall and B is the allowed minimum clearance with a protector, then the protector must be wide and high enough so that it protects all wall surface within distance A of the stove. In the example, the wall protector must be about 7 feet wide. Why this is probably excessive is explained in Appendix 2.

dicular projection of the fireplace opening on the wall to be protected).

Particular Wall Protector Systems and Materials

The three essential features for most wall protectors are:

1. Non-combustibility of all materials, including mounting or supporting systems.

2. A well-ventilated air space between protector and wall.

3. Adequate strength and rigidity so that the protector and the air gap will be durable.

Specifying particular systems in detail for protecting combustible walls and ceilings requires some personal interpretations and judgments, since not enough research has been done to prove or disprove the reasonable safety of many systems. NFPA has a list of approved methods (Table 2–5) but it has been criticized for not specifying if and how the 1-inch air gap is to be ventilated, not addressing how much support is necessary to assure the 1-inch gap is not lost

through denting or distortion or breakage of the wall-protecting material, not allowing systems with more esthetic appeal than sheet metal or asbestos millboard, and for including asbestos-containing protectors (asbestos fibers can cause cancer).

I do not generally recommend the use of any *non-ventilated* wall or ceiling protector, even those listed by NFPA and building codes. As explained in Appendix 2, the effectiveness of such protectors depends on the amount of insulation in the wall being protected. It is unsafe to rely on non-ventilated protectors for reduced clearances to insulated walls. On uninsulated walls such protectors are reasonably safe.

For all air-cooled wall protectors, adequate air movement is assured if all edges are left open, and this is recommended. However, in the case of a protector panel on a single flat wall away from corners, either both sides may be closed (circulation in at the bottom edge and out at the top), or the bottom may be flush with the floor (circulation in at the lower sides and out at the top). Wall protectors that cover two walls in a corner must allow input air flow all along the bottom edge. All ventilated wall protectors should be open along the top.

53

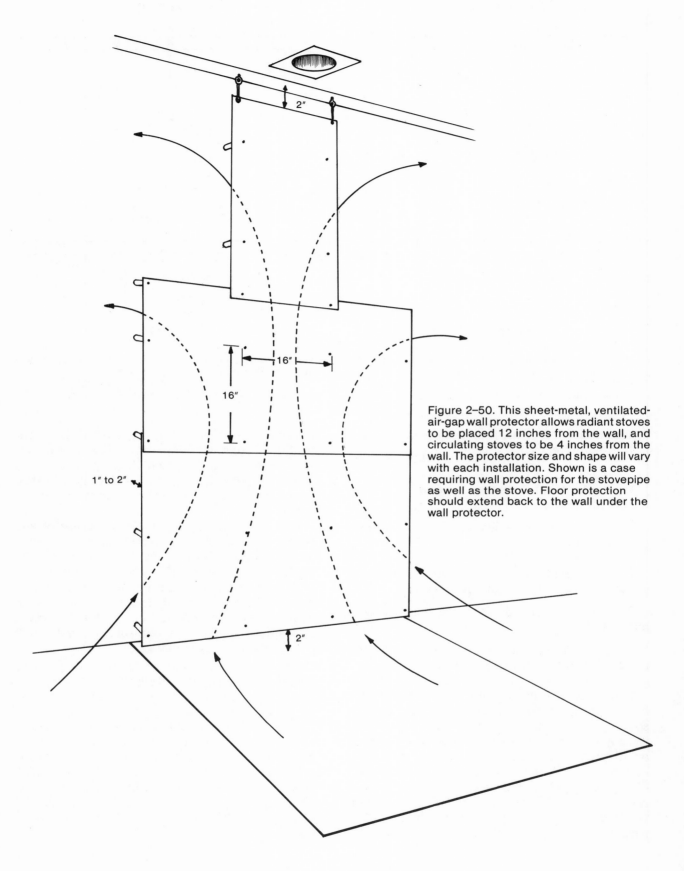

2"

16"

16"

1" to 2"

2"

Figure 2–50. This sheet-metal, ventilated-
air-gap wall protector allows radiant stoves
to be placed 12 inches from the wall, and
circulating stoves to be 4 inches from the
wall. The protector size and shape will vary
with each installation. Shown is a case
requiring wall protection for the stovepipe
as well as the stove. Floor protection
should extend back to the wall under the
wall protector.

54

Although a 1-inch air space behind ventilated wall protectors is generally considered adequate, a clearance of about 2 inches has two advantages. Accessibility is better for cleaning, inspection, and minor maintenance. In addition flow is less likely to be dangerously restricted if the panel warps slightly. Since reduced clearances are measured from the original wall surface, a wider air space does not require the wood heater to be installed any further out into the room. I recommend the 2-inch space.

I recommend the first two protection systems discussed below as being the best overall—thermally safe, mechanically strong, not made of asbestos, and using only commonly available materials.

1. Ventilated sheet metal (Figure 2–50). Use of 0.013 inch (28 gauge) sheet metal spaced out at least 1 inch is probably the least expensive wall protection system for reduced clearances. It permits radiant stoves to be placed 12 inches from a combustible wall, unless of course a side-exiting stovepipe requires more clearance than this would provide. Convenient hardware for mounting wall and ceiling protectors of sheet metal are porcelain (not plastic) electric fence or "line post" insulators, or short lengths of small diameter thin-walled pipe or tubing. The holes through their centers are convenient for nails or screws; small kiln-shelf stilts may also be used (see Appendix 5). If 1-inch spacers are used, they should be placed every 16 inches horizontally and vertically to ensure the air gap is not lost through warpage or denting of the sheet metal. If 2-inch spacers are used, they may be placed every 32 inches. It is sometimes recommended not to place spacers directly behind the stove or other especially hot spots to minimize the chance of heat conducting through the nail or screw to the wall, although the need for this has not been demonstrated. For the same reason, it may be wise to avoid thick-walled or large diameter spacers of high conductivity materials such as copper and aluminum.

Wood should not normally be used as spacers or for support. However, if the protector panel is as large as specified in Figures 2–48 or 2–49, wood can be used safely at the edges of the panel only; something else must be used for a spacer for the interior regions of the panel.

If the sheet metal needs extra support due to its weight, it can be hung from the ceiling with noncombustible materials. Various metal brackets and extrusions may also be useful as wall supports.

If bare galvanized sheet steel or sheet aluminum is unattractive, a possible remedy is high temperature paint. Although the infrared reflectivity[4] of the bare shiny metal helps keep the wall even cooler, it is not essential.

An alternative is to use bare copper, a metal many people find attractive. Any metal with any finish is safe enough, but the infrared reflectivity of bare clean copper and its thermal conductivity will keep the protected wall even cooler. Copper's high thermal conductivity allows much of the heat absorbed in the copper right behind the stove to travel away laterally in the metal. This spreading of the heat prevents any part of the sheet from getting very hot. If the copper is clean and bare, its high reflectivity will bounce back almost all of the heat readiated towards it back into the room; the copper then has little opportunity to get warm in the first place. I have found such wall protectors are rarely even warm immediately behind a nearby stove or stovepipe. Disadvantages of copper are its cost and its softness. Two dollars per square foot was a typical price in 1979 for copper sheet. Copper sheet bends and dents easily, which can mar its appearance. I would recommend 0.020 inch minimum thickness, or that the copper sheet be formed over some other noncombustible material for additional stiffness. Sheet steel or the better quality manufactured stove mats could be used (see item 6 following).

Large sheets of metal tend to warp when hotter in one area than another. The warping is unattractive and can cause the metal sheet to touch the wall. In addition to using enough spacers as previously discussed, constructing the protector of more than one slightly overlapping piece of sheet metal can reduce warping.

2. Ventilated brick wall (Figure 2–51). A 4-inch thick facing of mortared common brick, spaced out a minimum of 1 inch, provides safe protection for reduced clearances (12 inches for radiant stoves), while at the same time providing some heat storage and possible esthetic advantages. Gaps between bricks as shown in Figure 2–51 can provide air flow at the bottom. Particularly if the spacing is only 1 inch from the wall, the mortar overrun should be scraped off to leave a smoother inner surface for unobstructed air flow.

Since the weight of the brick and mortar is

---

4. Low emissivity on the back side of bare metal sheets also contributes to keeping the wall cool. Note that even a clear paint finish on bare metal will destroy its high infrared reflectivity and low infrared emissivity.

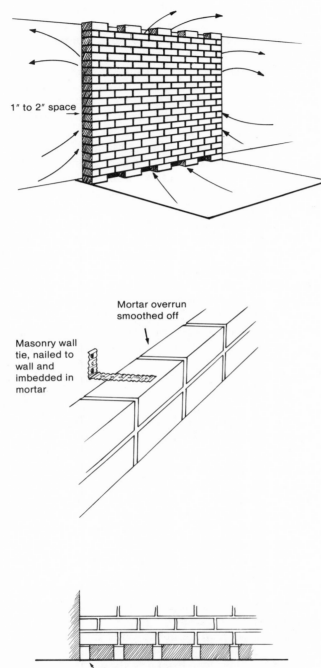

Figure 2–51. This air-cooled brick wall protector is not allowed by most building codes, but is safe. It must be tied to the house wall with masonry ties or the equivalent. The protector is shown extending to the ceiling. The stove size and stovepipe and chimney locations determine whether this is necessary. Floor protection should extend all the way back to the wall. Safe clearances are the same as for a ventilated sheet metal protector.

typically 1000 to 2000 pounds, such a structure should probably only be placed next to a bearing wall. Mechanical support between the brick and the existing wall is also important; ties should be placed about one per square foot, or about every 9 inches along studs on 16-inch centers, or otherwise if required by local codes.

3. Ventilated asbestos millboard (not recommended by the author, but approved by most codes). Only asbestos millboard is recommended by NFPA, although asbestos cement board, being more readily available, is often used in its place.

Asbestos millboard is a low-density, soft material that is easy to cut with a knife or saw. Asbestos cement board is a dense, hard and brittle material that chips easily and is more difficult to saw or break. The asbestos millboard, having a more porous structure, is a better insulator, but when used as a ventilated wall protector, this slight possible advantage becomes insignificant compared to the insulating or cooling effect of the ventilated air space itself.

The most significant difference between the two materials when used as ventilated wall protectors is that the cement board is more susceptible to cracking from mechanical or thermal shock. How frequently this happens in practice is not well established. I suspect that the cement board is not now included in most codes only because it was not chosen to be included in original tests on clearances and wall protectors conducted by Underwriters Laboratories in the early 1940s.

The major problem with these materials is the health hazard of asbestos fibers. Very little of either type of asbestos board is being made in this country because government health regulations have made it impractical to operate manufacturing plants. It is not known quantitatively whether the actual exposure to fibers from asbestos wall protectors is significant. Thus I am inclined to be cautious and not use asbestos materials.

If asbestos is used as, or as part of, a wall or floor protector, I strongly recommend painting the asbestos to help keep the fibers from coming loose. If asbestos board must be cut, don't inhale the dust. I recommend doing the work outdoors downwind from your and your neighbor's houses. A breathing mask should be used, and I even suggest holding your breath while cutting and then moving upwind before breathing again.

The required mechanical support and ventilation for asbestos board wall protectors is the same as for sheet metal protectors.

Note in Table 2–5 that according to NFPA the use of asbestos millboard to create a ventilated air space protector does not permit placing the stove or stovepipe as close to a wall as does use of the other materials. This is somewhat arbitrary; a reduced clearance of 12 inches is reasonably safe with *any* ventilated wall protector. The experimental data on which NFPA's recommendations are based show very similar wall temperatures for air-ventilated asbestos board and air-ventilated sheet metal protectors. For simplicity, NFPA uses only a few discrete reduced clearances (30, 24, 18, 12 inches, etc.). The only slightly worse performance of the asbestos board resulted in its being pushed into a clearance category 50 percent different.

4. *Masonry veneer.* Thin (½- to 2-inch) masonry veneer materials such as Z-Brick, Real-Brick, Magic Brick, and various precast stone veneer panels are useful principally for their appearance, not for their protective value. They are actual brick or stone-like materials which come either as individual "tiles" or as one-piece slabs looking like a piece from a mortared masonry wall. These veneer materials are intended to be applied directly to an existing wall for cosmetic effects only. This use does not permit reduced clearances; although noncombustible, these materials have very little insulating value and do not keep a wall cool.

When spaced out an inch or more from a wall, a masonry veneer can create a ventilated gap that permits reduced clearances. The problem is one of mechanical strength. The tiles must be securely mounted on something strong, noncombustible, and compatible as temperatures rise and fall repeatedly. Testing is needed to verify some practical support schemes. It may be satisfactory to apply the veneer over metal lath screwed to sheet metal. Mortar, not glue, must be used since most glues are combustible and fail at high temperatures. Obviously considerable mechanical support would be necessary to hold up the weight, to ensure the one-inch air gap, and to prevent stress or flexing that would crack the masonry. Since the sheet metal itself constitutes an adequate ventilated wall protector, covering it with a masonry veneer would be desirable only for its appearance.

One-piece, metal-reinforced thin masonry panels can be strong enough and are becoming available. More than one panel may be necessary for adequate coverage. The weight of such panels may be coming down due to development of lightweight concrete/insulation mixtures.

There are panels that look like stone or brick but are actually made mostly of wood or wood fibers, fiberglass and resin or gypsum board. These are all dangerous to use as wall protectors.

5. *Gypsum board, Sheetrock, wallboard, or plasterboard.* I do not recommend materials such as gypsum board to space out from a wall to provide a ventilated air space.

United States Gypsum does not recommend use of ordinary gypsum board at temperatures above about 130° F. It contains moisture which is critical for its strength. At temperatures above about 130° F., this water is gradually driven out and the plaster crumbles. Wall protectors can easily reach this temperature.

The fact that a gypsum board or any other material has a "fire rating" of some kind is not relevant. Fire ratings concern the rate at which fire can spread through a material or construction and do not imply that the material will survive at high temperatures, or even that it is noncombustible.

6. *Stove mats (floor mats).* Good-quality stove mats or floor protectors can be used to construct ventilated wall protectors. Two possible problems exist. First, traditionally stove mats are constructed of asbestos millboard covered with sheet metal. Although this construction is certainly noncombustible, it contains asbestos, a material many people prefer not to have in their homes. Second, stove mats are now being constructed out of different materials whose noncombustibility is difficult for the consumer to determine. In some units cardboard or wood particle board is used. Flame-retardant chemicals may be incorporated into such materials, but this may still be inadequate. Neither "flame retardant" nor "fire rated" implies noncombustibility. Unfortunately there is not yet a generally recognized performance standard for stove mats; hence there is no standard testing and listing of such products.

If stove mats are used, more than one may be necessary to achieve the required coverage. Cracks of ¼-inch or less, or small holes of 1 square inch or less are not serious as long as wider gaps and holes are not likely to develop by the panels shifting.

NFPA lists the traditional stove mat construction as being suitable for some degree of wall protection with no air gap. The 0.013-inch (28 gauge) sheet metal on ¼-inch asbestos millboard applied directly to a wall permits the normal 36-inch stove clearance to be reduced to 18 inches. With a 1-inch ventilated air gap, the clearance may be 12 inches.

I strongly recommend providing a ventilated air gap even when not required by NFPA, for two reasons. I feel that easy inspection of combustible surfaces is important as a general safety principle. I feel more comfortable if I can periodically check with a flashlight for wall discoloration than if the wall is "permanently" covered and hence hidden from view. Of course if the wall protector is adequate and proper stove-to-wall clearances are observed, this inspection should not be necessary. But it happens frequently enough that both materials and installation details differ from the exact wall protection schemes tested and approved by safety authorities that easy inspection is useful. In addition, if the wall behind the protector does overheat, the odor will quickly enter the living space of the house, giving early warning of a dangerous situation.

7. Insulation. NFPA lists three forms of wall protection for reduced clearance that do not require a ventilated air space. One uses sheet metal on asbestos millboard, discussed above. The other two use sheet metal or asbestos millboard backed with insulation (Table 2–5, entries e and f, and Figure 2–52).

As explained in Appendix 2, I believe that protectors that are not air-ventilated do not always give enough protection to justify NFPA's reduced clearances. This is particularly true of the insulated-type protectors applied to an insulated wall. Some heat conducts through all types of protectors. In ventilated types, this heat is carried away in the air flow. In unventilated types covering uninsulated walls, the heat safely conducts through the wall. But if the wall is insulated, the heat cannot get away easily and the temperatures rise immediately under the protector at the surface of the wall. Thus I strongly recommend against unventilated protectors on insulated walls and suggest using only ventilated protectors in all cases.

Whenever insulation is used as a protector, it must of course be noncombustible and must stay in place. The sheet metal or asbestos board covering required by NFPA is partly to give the insulation more mechanical strength and durability. NFPA specifies, in NFPA Standard No. 211, insulation with a minimum density of 8

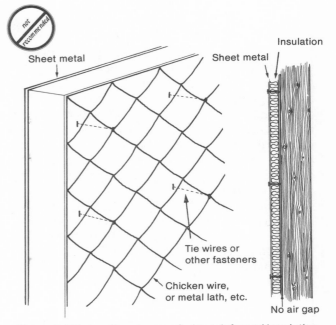

Figure 2–52. A wall protector of wire-reinforced insulation covered with sheet metal. The wire reinforcement is to hold the insulation in place. NFPA says this protector enables radiant stoves to be placed 12 inches from the protected wall, and circulating stoves 4 inches from the wall. I believe this protector is not always effective, particularly when placed on an insulated wall.

pounds per cubic foot and a minimum melting point of 1500° F. I recommend ceramic wool insulation, such as Fiberfrax (see Appendix 5). Ordinary fiberglass insulation can be used, but it must not have a paper or foil backing as both these backings are combustible. In addition, most fiberglass insulation uses a very thin layer of glue or binder to hold fibers together where they cross. When heated excessively, the glue will smoke away. This causes the insulation to be less springy and resilient. As a result it may compress, losing some of its insulating value. For this reason, and to come closer to NFPA's recommended density, when fiberglass insulation is used it should be moderately compressed—for instance, a 3½-inch bat might be used for a 1-inch layer of insulation. Typical residential fiberglass insulation has a density of a little less than 2 pounds per cubic foot.

The *metal-reinforced* insulation specified by NFPA is not readily available. The objective of the reinforcement is to assure the insulation stays in place without slipping, sagging or separating. This objective can be met by sandwiching unreinforced insulation between sheet metal and chicken wire using wire ties running through the insulation (Figure 2–52).

8. Thick masonry. Masonry veneer requires a ventilated air space behind it to constitute protection of a combustible wall for reduced stove clearances. Most authorities believe a 4-inch masonry facing would make safe a 24-inch clearance between a radiant stove and a combustible wall. What thickness of unventilated masonry is equivalent to a simple ventilated sheet metal protector, permitting a 12-inch clearance for a radiant stove? This question has not been answered experimentally. A brick wall would have to be 4 to 6 feet thick to have the same insulating value as an ordinary fiberglass-insulated 3½-inch thick wall. Thus since according to NFPA 1 inch of insulation is adequate protection on an *uninsulated* wall for a 12-inch clearance, 1 to 2 feet of masonry is probably adequate in this case. The high transverse (sideways) conductance of heat storage capacity of thick masonry might permit this thickness to be reduced.

In practice such thick masonry structures would be too heavy for most floors to support safely, so virtually all practical wall protectors constructed of masonry materials need a ventilated air space behind them. It is also apparent that an existing ordinary solid wall of brick, stone, or concrete should not be considered safe for clearances less than about 24 inches to radiant stoves unless there is no combustible material in the wall. Thus, for example, solid brick or concrete walls with foam insulation or with wood siding on the outside may need additional protection for clearances less than about 24 inches (Figure 2–53). But again, the sideways conduction of heat in the masonry, as well as their high heat storage capacity, may make safe much smaller clearances. More research is needed.

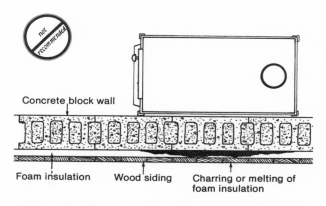

Figure 2–53. What constitutes a non-combustible wall? Stoves should not be placed against masonry walls with combustible materials on the other side, as this view from above illustrates.

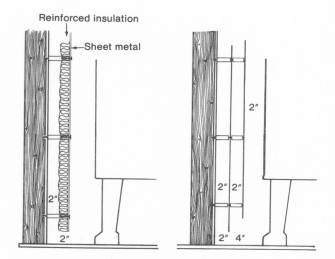

Figure 2–54. Two doubly effective wall protectors, allowing radiant stoves to be placed 6 inches from the protected wall. This is my estimate; NFPA and building codes do not recognize this kind of protector. Brackets or ceiling hangers may be used to support it.

9. Double or combination protection systems. Ventilated-air-space wall protectors allow clearances of 12 inches between a radiant stove and the original wall surface (unless minimum stovepipe clearances require the stove to be further away). Less clearance is rarely needed but can be achieved. Since the degree of protection provided by following methods has not been measured, the minimum safe clearance is not known. It is likely to be significantly less than 12 inches; I would guess 6 inches would be safe. This is consistent with NFPA's reduced clearance of 4 inches for a circulating-type stove (which has one built-in ventilated air space) from a wall with a sheet metal protector spaced out 1 inch.

One method has already been discussed under item 6—Insulation. It is a sheet metal exterior backed by 1 inch of reinforced insulation, mounted with a 1- to 2-inch ventilated air space behind it (Figure 2–54). If the stove is placed only a few inches away from the wall protector, I recommend using a higher temperature insulation than ordinary building insulation fiberglass, which is not recommended for use above 350° F. (See Appendix 5 for sources.)

Another method is to use panels to provide two independent ventilated air spaces. Care must be taken to assure adequate air flow through both spaces (Figure 2–54).

The closer protector panels and wood heaters are to each other, the hotter the panels may be. One consequence can be more warpage of sheet metal panels, and cracking of masonry. To mini-

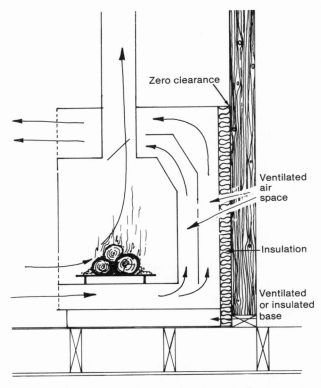

Figure 2–55. A zero-clearance fireplace. Its back can be placed safely against a wood wall. The exterior of the unit is only warm to the touch because of the two ventilated air spaces plus the layer of insulation.

mize sheet metal warpage, secure and frequent (about every 16 inches) spacers are important. Making the panel out of more than one piece of sheet metal can help. Spacers should not be larger in diameter than necessary for their structural functions in order to minimize heat conduction through them.

This protection approaches that used in "zero-clearance" metal fireplaces. The back of these fireplace units can be placed safely in direct contact with a wood wall. Here the fire is only 4 to 6 inches from the wall. The wall protection is of course built into the appliance, but the principles are the same as for the wall protection systems discussed above. Two ventilated air spaces and one insulated space are used in some zero-clearance fireplaces, as illustrated in Figure 2–55.

If you want to experiment with clearances, I suggest acquiring a device to measure the temperature of the wall surface behind the appliance. If in the case of very long lasting and very hot fires, wall temperatures do not exceed 200° F., the system is probably safe. Alternatively, if you cannot hold your hand pressed against a wood or plaster surface, it is probably dangerously hot.

Reduced Clearances for Stovepipe

For unprotected combustible walls and ceilings, the minimum clearance for stovepipe is 18 inches, according to NFPA. As is the case with large stoves, these standard clearances are not for continuously red-hot operation. Reduced clearances are safe if the wall or ceiling is appropriately protected. The principles and methods of protection are similar to those for the wood heaters themselves, a ventilated air space being the most reliable. All but the masonry constructions are appropriate for ceiling protection. The safe reduced clearance for each type of protection as established by NFPA is indicated in Table 2–5. Table 2–6 contains my recommendations.

For wood heaters with stovepipe connections to

General Case

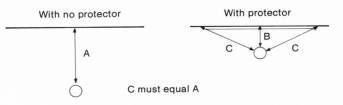

With no protector   With protector

C must equal A

An Example:

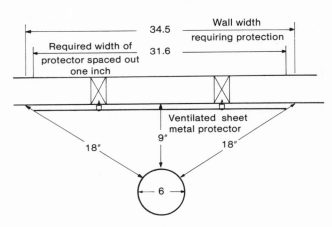

Figure 2–56. Widths of wall and ceiling protectors for stovepipe, according to NFPA and most building codes. The clearance of stovepipe to unprotected walls and ceilings (A in figure) is 18 inches for any size pipe. I recommend as safe a 24-inch width for wall and ceiling protectors for all stovepipe diameters up through 8 inches and for all reduced clearances covered in Tables 2–5 and 2–6.

the side or rear of the appliance, minimum clearances between the stove and a wall will usually be determined by minimum stovepipe clearances if the appliance is oriented with stovepipe collar towards the wall.

The NFPA-recommended extent of the protection is such that no unprotected part of the wall or ceiling is within the unprotected clearance of the stovepipe (Figure 2–56 and Table A6–3 in Appendix 6). This NFPA recommendation leads to unnecessarily large protectors, in some cases more than 3 feet wide. In fact a 24-inch protector width is probably adequate for any stovepipe diameter up through 8 inches and for any reduced clearance that is allowed in Tables 2–5 or 2–6.

If stovepipe rises vertically from the wood heater and then exits through the near wall, the wall protection need only be as high as the vertical portion of the stovepipe. If the pipe rises to a chimney at the ceiling, the wall protector should extend to 2 inches from the ceiling, the space being for air circulation. Gaps in the protector panel at locations such as this are not dangerous unless the reduced clearance is very small—considerably less than 9 inches. Although radiation from the stovepipe can reach the wall directly, the intensity or flux of the radiation can at most be only about half of that under more common circumstances because the radiating pipe only extends down, not up, from ceiling level.

Ceiling, wall, and floor protection extents and equipment clearances are summarized in two particular examples in Figure 2–57.

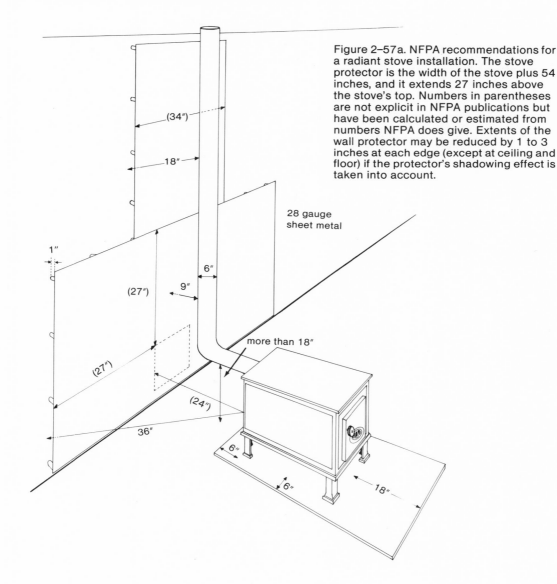

Figure 2–57a. NFPA recommendations for a radiant stove installation. The stove protector is the width of the stove plus 54 inches, and it extends 27 inches above the stove's top. Numbers in parentheses are not explicit in NFPA publications but have been calculated or estimated from numbers NFPA does give. Extents of the wall protector may be reduced by 1 to 3 inches at each edge (except at ceiling and floor) if the protector's shadowing effect is taken into account.

28 gauge
sheet metal

(34″)

18″

1″

(27″)

6″

9″

(27″)

more than 18″

(24″)

36″

6″

6″

18″

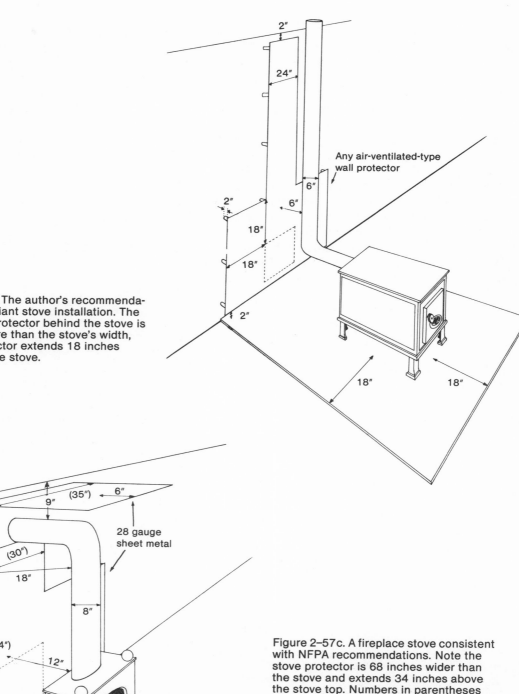

2″
24″
Any air-ventilated-type
wall protector
6″
2″
6″
18″
18″
2″
18″   18″

Figure 2–57b. The author's recommendations for a radiant stove installation. The
width of the protector behind the stove is
36 inches more than the stove's width,
and the protector extends 18 inches
higher than the stove.

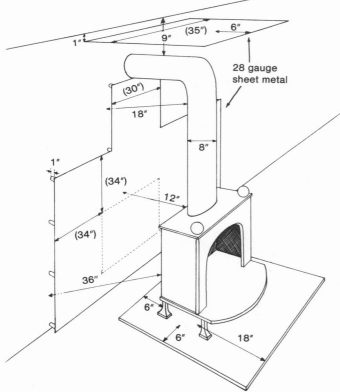

(35″)   6″
1″   9″
28 gauge
sheet metal
(30″)
18″
8″
1″
(34″)
12″
(34″)
36″
6″
6″   18″

Figure 2–57c. A fireplace stove consistent
with NFPA recommendations. Note the
stove protector is 68 inches wider than
the stove and extends 34 inches above
the stove top. Numbers in parentheses
are not explicit in NFPA publications but
have been calculated or inferred from
NFPA guidelines. Extents of the
protectors may be reduced 1 to 3 inches
at some edges if the shadowing effect of
the protector is taken into account.

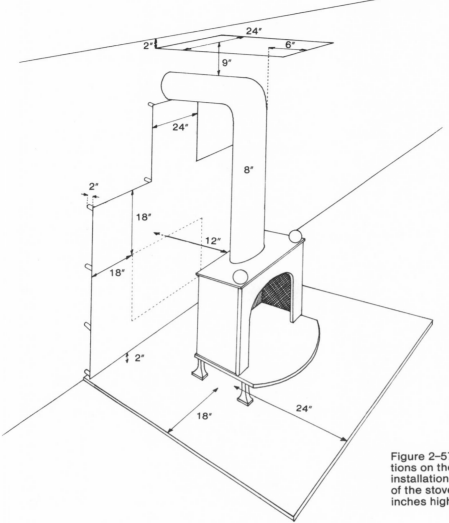

Figure 2–57d. The author's recommendations on the same fireplace stove installation. The wall protector is the width of the stove plus 36 inches, and is 18 inches higher than the stove.

Also, the floor *beside* stoves receives large amount of radiation from the *sides* of the stove. In fact for stoves with relatively short legs, radiation from the sides can make the floor hotter than portions of walls for which NFPA recommends substantial protection (see Appendix 2 and Figure A2–10).

Floor Protection

Unlike walls, combustible floors must be protected against both overheating by radiation and burning material such as sparks and coals falling or rolling onto the floor.

Normally only small pieces of burning material land on the floor. Fireplace stoves should have andirons, a raised edge on a grate or extended hearths should prevent burning logs from getting out of the fire chamber onto the floor. Thus with reasonable care in the operation of stoves and fireplace stoves only sparks and small coals will usually fall out, and this mostly during refueling and ash removal. NFPA recommends only sheet metal, 24 gauge or thicker, for protection against this hazard. A floor covering material for protection from sparks and small glowing coals should be noncombustible, continuous (no cracks or holes), and sufficiently strong not to crack, tear, or puncture with normal use.

Floor protection against radiation from the stove may require considerably more than sheet metal. Stove bottoms are almost always cooler than stove sides. Ash in the stoves insulates the bottom from the fire, and flames naturally move

63

upwards away from the bottom. In addition, many stoves have either firebrick-lined bottoms or ash drawers under grates. Cats often sleep comfortably under operating stoves.

However, some stove bottoms can get very hot. Hot bottoms are more likely on heavy, thick-walled stoves because heat can more easily conduct down through the thick walls into the bottom.

NFPA's recommendations for protection of combustible floors under stoves and fireplace stoves are as follows (NFPA HS-10):

1. If there is an 18-inch open air space between the bottom of the stove and the floor 24 gauge or thicker sheet metal gives adequate protection.

2. If there are between 6 and 18 inches of open air space, the floor protection material should be ¼-inch asbestos millboard covered with 24-gauge sheet metal.

3. When there is less than 6 inches of open air space, the floor should be protected with 4-inch thick hollow masonry units arranged with the holes interconnecting and open to allow free air circulation through the floor protector (see Figure 2–58). The hollow masonry should be covered with 24 gauge sheet metal.

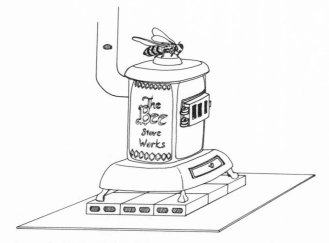

Figure 2–58. A ventilated floor protector recommended for stoves with legs less than 4" long or unusually hot or large bottoms. The holes through the 4-inch masonry blocks must be aligned and unobstructed, to permit air flow through the protector. Mortar should not be used. If the ventilated masonry protector is used only directly under the stove itself some other suitable floor protector should be used as an underlying base, and should extend out to the normal distances for floor protectors. If the masonry is used as the only protector it should have sheet metal either under or over it to prevent small sparks falling through the cracks to the floor. The blocks should be kept close together, but the exposed holes at the sides must not be covered, as with a retaining frame.

4. As always, if a stove is listed, it is to be installed (including floor protection) in accordance with the manufacturer's instructions.

NFPA's recommended extents for floor protectors are:

18 inches beyond any side with a door in it,
6 inches beyond all other sides

My feeling is that NFPA's guidelines on floor protection are in some cases inadequate in terms of extent, and that a wider variety of materials should be permitted.

NFPA's 18-inch extent beyond sides with doors is reasonable for *stoves*, but perhaps the 18 inches should be redefined as extending to the sides of the door opening as well as to the front since glowing coals often land or roll to the sides. For *fireplace stoves* 18 inches can be inadequate. Radiation from the fire with the doors open can overheat a floor 18 inches away. For radiant-type gas heaters, NFPA requires floor protection to extend 36 inches in front. Thus for medium and large fireplace stoves, I recommend extending the floor protection out to at least 24 inches beyond the fireplace opening. I also recommend this extent for any fireplace stove used extensively with doors open without a spark screen. These same considerations apply to traditional fireplaces and prefabricated metal fireplaces (Chapter 5).

NFPA's floor protector extent beyond sides without doors is 6 inches. This may be inadequate for large, short-legged, or hot stoves, due to radiation from the stove's side heating the floor. There is not yet much evidence as to how much wider floor protectors should be. Some state building codes require a 12-inch extent beyond sides without doors. As is true with all protector and clearance situations, stove size and surface temperature are critical. I believe that the 12-inch floor protector extent is adequate in the vast majority of cases, but that 18 inches may be advisable as a general recommendation to cover virtually all stoves, at least until more evidence is available. Stoves with inherently cooler sides, such as stoves with firebrick liners, will not always need even 12 inches. For circulating stoves, a 6-inch side extent is adequate.

Whenever a stove is installed with reduced clearance to walls, the floor protection should extend all the way to the wall, particularly if the stovepipe connects to the side of the stove facing the wall. The more confined nature of the space

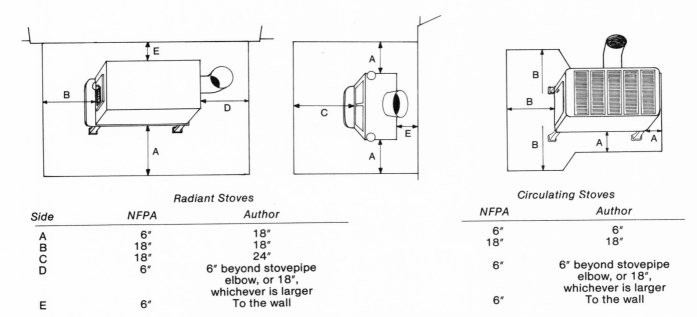

Radiant Stoves

Side	NFPA	Author
A	6"	18"
B	18"	18"
C	18"	24"
D	6"	6" beyond stovepipe elbow, or 18", whichever is larger
E	6"	To the wall

Circulating Stoves

NFPA	Author
6"	6"
18"	18"
6"	6" beyond stovepipe elbow, or 18", whichever is larger
6"	To the wall

A (any side except where E applies)   B (sides with fuel-loading or ash-removal door)   C (sides where open fire is an intended use)   D (sides with side-connected stovepipe) if the stovepipe is less than 18 inches from the floor, additional protection for the floor is needed   E (sides with reduced clearance to the protected wall)

Figure 2-59. Minimum recommended extent of floor protection according to NFPA, and according to the author. All dimensions are measured horizontally from stove sides. If a *ventilated* masonry floor protector is required, it need only cover the floor directly under the heater (including its legs) if another protector is used as a base and has the full extent required in the figure above.

and the possible additional radiation from the stovepipe make this additional floor protection desirable. In all cases of side-connecting stovepipe, the floor protector should extend about 6 inches beyond the pipe elbow.

Just as stove type does not enter into NFPA's recommended floor protector extents, neither is it considered for floor protector materials. For the sake of simplicity this may be desirable. However, there are stoves that I feel have inherently cooler bottoms and sides and thus require less floor protection: circulating stoves with grates and an ash drawer, and circulating stoves with an open, air-gapped (1 inch or more) metal shield underlying the entire stove bottom as a part of the stove. In some designs stronger ventilation of this air space is obtained by connecting it to a similar air space on the back of the stove; the rising hot air on the back pulls room air in under the bottom. My belief is that circulating stoves with either a grate with ash drawer, or a ventilated bottom shield, if they have 4-inch or longer legs, require only sheet metal floor protection (Table 2–7). (In NFPA's Fire Protection Handbook, 14th Edition, this floor protector is permitted for *any* stove with 4-inch or longer legs.)

NFPA's selection of materials for floor protectors is limiting. Plain sheet metal is usually unattractive, easily bent, and hazardous around the edges. Sheet metal over asbestos may constitute a health hazard of a different sort due to release of asbestos fibers into the air. (The actual exposure has not been assessed, so the seriousness of this hazard is not known.) Hollow masonry blocks are also unattractive to many people.

There are some alternative materials and modifications, some of which offer the same or more protection, and some of which may not. The traditional manufactured stove mat or stoveboard floor protector is sheet metal wrapped over asbestos board. There are available today a number of other manufactured stoveboards. Some have combustible cores made of cardboard or wood fiberboard. Even if treated with flame-retardant chemicals, these materials often can still smoke and burn. Thus they should not be used where sheet metal over asbestos is required, at least until some testing has been done. They of course may be used where only sheet metal is called for. One brand of cloth-like floor covering sold as a noncombustible floor protector for in front of fireplaces in fact burns very easily.

Table 2-7. Floor Protection (Author's Recommendations)[1]

Stove Design	Minimum Combustible Floor Protection
Any stoves with 18″ legs or longer. Circulating stoves with 4″ or longer legs *and either* grates and ash drawer *or* exterior ventilated baffle on stove bottom, with 1″ minimum gap	Sheet metal, 24 gauge. A decorative masonry overlay is OK.
Stoves with 6″ to 18″ legs without ash drawer or ventilated exterior bottom baffle	4″ of masonry or 2″ of sand or small gravel with 24-gauge sheet metal either underneath or on top (See Figure 2–60.)
All other stoves	4″ hollow masonry with holes aligned and open for air circulation, covered or underlain with 24-gauge sheet metal

All floor protectors must be spark tight. Thus protectors constructed of unmortared masonry units, sand, or gravel need a sheet metal base or cover.

Leg length here means distance from stove bottom to the protector's surface under stove. For stoves with exterior bottom baffles, leg length is measured from the baffle to the floor protector.

Stove legs should not penetrate through any part of the floor protector but must rest on the protector's top surface.

The 24-gauge sheet metal is about 0.024 inch thick. The major importance of thickness is reasonable mechanical strength, but greater thickness and use of higher thermal conductivity materials results in more transverse conductance which can help spread heat away from a potential hot spot.

For an extra margin of safety, remove any insulation that is in the floor directly under the stove.

If a heater is often operated so hot that its exterior surface glows red hot, more protection may be required. Red-hot operation is always damaging to the heater and is not recommended.

Exceptionally large stoves with the surface area of any side or the bottom exceeding about 1000 square inches may need extra protection.

1. These clearances and materials for protecting combustible floors under stoves and fireplace stoves are not necessarily approved by NFPA or building codes.

If the core of a sheet metal-wrapped floor protector is truly noncombustible and has an insulating value equivalent to ¼-inch asbestos millboard, these stoveboards may be used where the asbestos millboard/sheet metal protector is specified.

If more than one stoveboard is used to obtain the desired extent, either the whole protector should be underlain with sheet metal or the boards should be arranged so there is no junction between adjacent boards in the area in front of stove doors, where sparks and hot coals may fall. The objective here is to prevent fires due to burning material in the crack or space between adjacent stoveboards.

Masonry materials such as stone and brick can be used as part of floor protectors, with a decorative and protective function. Added masonry never decreases the protective value of a floor covering. Thus any kind of masonry may be laid loosely or set in mortar, on top of sheet metal. The result may be used where sheet metal alone would suffice. Similarly, stoveboards may be overlain with masonry units.

Under what circumstances added masonry or masonry alone starts adding to the protective value is not certain until more testing has been done. However, I expect that 4 inches of solid masonry or 2 inches of sand or small gravel (Figure 2–60) offer protection at least as effective as the traditional metal-plus-asbestos-millboard stoveboard.

Some interesting stoveboards are becoming available made of lightweight aggregate concrete, in some cases with metal reinforcing. These units can be moderate in weight and possess significant insulating value. They can also have esthetic

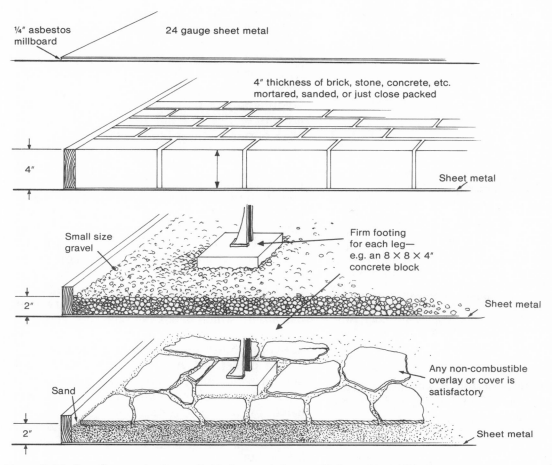

¼" asbestos millboard

24 gauge sheet metal

4" thickness of brick, stone, concrete, etc. mortared, sanded, or just close packed

4"

Sheet metal

Small size gravel

Firm footing for each leg— e.g. an 8 × 8 × 4" concrete block

2"

Sheet metal

Sand

Any non-combustible overlay or cover is satisfactory

2"

Sheet metal

Figure 2-60. Four floor protectors suitable for most stoves with legs at least 6 inches long, according to the author. NFPA and most building codes recognize only one of these protectors, the asbestos millboard plus sheet metal. I believe the other three offer at least as much protection, as explained in Appendix 2. They also have the advantages of not containing asbestos and of offering more esthetic variety.

exteriors. Ventilation can also be incorporated into the panel. Again, testing is needed to establish the degree of protection these units may offer.

The ultimate in protection is provided with a ventilated air space (Figure 2–58). When ventilated masonry is called for, it is only required by NFPA to be directly under the wood heater. However, protection equivalent to the asbestos millboard/sheet metal combination should be provided out to the normal extent beyond the unit's sides (Figure 2–59).

Removing any insulation in the floor directly under a wood heater will provide an extra margin of safety regardless of floor protector type.

Room Size

Stoves should not be installed in tiny rooms. Even if all clearances and normal protections are

observed, a small closed, insulated room can get dangerously hot. The clearances and wall protection schemes all presume that the average temperature in the room is not abnormally high. In practice the problem of too small a room may arise if one is trying to make a stove into a furnace by placing it in a closet and ducting air through the closet. If the air flow should ever be blocked in a situation like this, very serious overheating of the room could occur. Most codes do not address this problem as it relates to wood heaters, and so common sense is the only guide. If you want to make a stove into a hot air furnace or circulating type heater, do not install the stove in an existing small room but build an enclosure around the stove using only noncombustible materials.

Another small-room application of wood heaters is in saunas. Again, a great deal of caution is merited since few if any wood heaters have

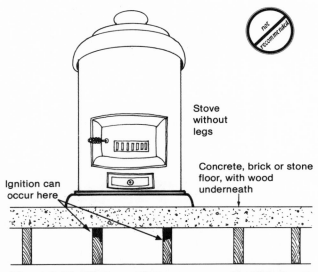

Figure 2–61. What constitutes a non-combustible floor? For an extreme case of a stove with no legs, even 4 inches of masonry may not protect wood underneath. In this case a ventilated floor protector should be used. If such a legless stove were installed on a combustible floor, two courses of ventilated blocks plus sheet metal should be used.

automatic overheat control. I suggest careful operation and/or a modified thermostatic control (for a higher temperature range).

Smoke Detectors

Smoke detectors in working condition should be part of every wood heating system. As a minimum, I recommend two—one in the room with the wood heater to warn of possible problems

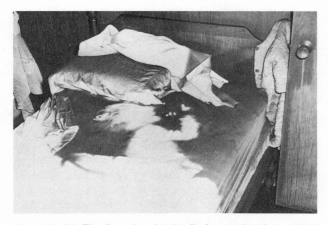

Figure 2–62. The "smoke-shadow" of a small girl who died of asphyxiation along with two sisters, a brother and her mother. Such accidents are much less likely in homes with smoke detectors. Photo courtesy of City of Willoughby, Ohio Fire Department.

there before the situation is beyond easy control, and one near bedrooms to save lives in any case.

There are some choices in selecting smoke detectors: battery-operated versus plug-in, and ionization versus photoelectric principles of operation. I have heard no compelling arguments favoring any of these choices. The plug-in units are susceptible to being unplugged and will not work during a power failure, whether fire-related or not. They also will not work if the fire damages the circuit they are using, but such occurrences are relatively rare. On the other hand, the battery units do not perform if one neglects or forgets to replace the battery, required about once a year. The photoelectric type is more sensitive to smoke and gives earlier warning of fires starting with smoldering conditions. The ionization type is more sensitive to invisible products of flaming combustion and may give earlier warning of fires starting with flames. The photoelectric type also has a bulb which needs periodic replacement.

Overall, the choice of which smoke detector to use is much less critical than the decision to use smoke detectors.[5] All types work satisfactorily and provide substantial reduced risks of fire damage, injuries, and deaths. The units should be listed by a major testing laboratory, such as Underwriters Laboratories, Inc. (UL), and should be installed according to the manufacturer's instructions.

Installation of Stoves in Fireplaces

Installing a stove in or in front of a fireplace can save the considerable trouble and expense of installing a new chimney. Three types of installations in masonry fireplaces are shown in Figure 2–63. Special cautions concerning installations in prefabricated fireplaces are discussed at the end of this section.

There can be some performance problems with stoves installed in front of fireplaces, due to two basic causes.

1. Typical stove installations have 4 to 7 feet of exposed stovepipe, which contributes 15 to 30 percent of the total heat output. Most of this potential heat gain is lost when stoves are installed immediately in front of a fireplace and vented up the fireplace throat. This is especially true of exterior fireplace chimneys, whose heat loss con-

---

5. Consumers Union has studied smoke detectors and made recommendations. See "Smoke Detectors," Consumers Report 41 (Oct. 1976) pp. 555–559.

ducted through the chimney walls is not a gain to the house.

2. Most fireplace chimneys are very oversized for most stoves (see Table 2–1). As a consequence there is often excessive cooling of the smoke due to the large surface area and low flue-gas velocity, and condensation, creosote buildup, and low draft can result. Three fireplace installation options are illustrated in Figure 2–63. Option A usually requires cutting a hole through the chimney above the fireplace throat. See Figure 2–32 for connection details. In option B the fireplace damper is opened or removed and the pipe is inserted up into the chimney. The damper opening is sealed as tightly as possible around the stovepipe using sheet steel and/or high-temperature insulation, such as ceramic wool (see Appendix 5). It is best to have a few feet of stovepipe extending up into the chimney. In option C the smoke is dumped in the fireplace chamber itself through a hole in a cover plate, usually a steel plate, over the fireplace opening.

Of the three installation options, A is the best. The exposed pipe rising directly up from the stove results in higher energy efficiency and better draft than either of the other two installations.

Installation A also has the advantage that the chimney is easy to inspect for creosote, and to clean when necessary. One only has to open the damper for good access. In installations B and C, the whole installation must be taken apart to get the same access to the chimney. This can be so much of a chore that it will discourage inspection and cleaning, resulting in a significant safety hazard.

Option C is usually the least satisfactory. It has lower energy efficiency than A and less draft and more creosote than both A and B.

In all three cases performance is less satisfactory if the fireplace and chimney are built into an exterior wall of the house. The cooler environment results in more creosote and less draft. Installation C can suffer particularly. If slow-burning fires are the rule, liquid creosote may flow out underneath the cover panel into the house. There have also been reports of solid creosote building up 2 or 3 inches thick inside the fireplace.

Some exterior fireplace chimneys are susceptible to flow reversal. I do not recommend venting a wood stove into a fireplace chimney which when cold has outdoor air flowing down it into the house.

Most of the possible creosote and draft problems can be lessened by decreasing the flue size.

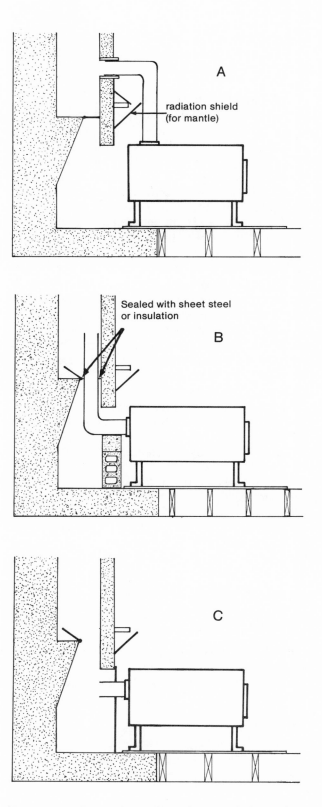

Figure 2–63. Three ways to install a stove in front of a fireplace. A will generally perform the best, yielding more heat output and better draft than B or C, and less creosote condensation than C. In most cases added floor protection beyond the fireplace hearth is necessary.

69

A new smaller flue or liner can sometimes be installed inside the large chimney. Stainless steel stovepipe is relatively durable and easy to work with.

Wood mantles and trim around fireplaces, and the wall in which the fireplace is built, may require protection if they are close to the stove (see Figure 2–44 to estimate how close is too close) or within three stovepipe diameters of the stovepipe. If this is difficult, shortcutting clearances to *exposed* wood such as mantles and fireplace trim is less dangerous than shortcutting clearances to *covered* wood such as studs under wallboard since deterioration is likely to be readily visible before it becomes dangerous.

The creosote problem can be even worse with prefabricated fireplaces. Some metal chimneys used with these fireplaces are of the thermosyphon type. The cooling of the flue can result in more creosote than in masonry chimneys. In addition, in some designs the creosote can drip down into the walls of the fireplace and apparently even onto the floor. Subsequent use of the fireplace as originally intended can ignite the creosote in the walls where only cooling air was meant to be. If the installation incorporates a cover plate over the fireplace opening, there is also danger of blocking openings for cooling air needed for safe use of the chimney.

I cannot recommend hooking a stove to a prefabricated fireplace, unless either the chimney is of the insulated type (see Figure 2–11) or a new stainless steel liner can be inserted the full length of the chimney and be connected directly to the stove or its steel stovepipe. This should both reduce the volume of creosote and keep it out of the fireplace inner structure.

Basement Installations

Installing a stove in a basement will save space upstairs, will confine the messiness of wood heating, such as bark, sawdust, ashes, and beetles, downstairs, and usually provides easy access to a chimney.

There can be problems. Placing a stove in a basement is not the most efficient way to heat the upstairs. Sixty to 80 percent of the heat output of a radiant stove is in the form of radiation, and radiation travels in straight lines until absorbed by a solid surface. Thus much of the stove's heat output will be absorbed by the basement floor and walls, and only some of this heat will ever find its way into the living spaces of the house.

In addition, not all the convected hot air heat at the ceiling will get upstairs since whenever the basement ceiling is warmer than the basement floor, heat is being transferred by radiation from the ceiling to the floor. The effect can be large. For a 500-square-foot ceiling at 75° F. and floor at 55° F., the radiant heat loss downwards from the ceiling to the floor is about 10,000 Btu per hour.

Thus one can expect to burn more wood in a basement-installed stove to get the same amount of heat upstairs compared to the same stove installed upstairs. If your fuel wood is free, or if you really like a warm floor, or if the stove is for emergency heating only, or if the basement is the only easy place to install the stove, or if heating the basement is a principal objective, basement installations make sense.

Another potential problem is chimney safety. One reason why many people consider basement installations is the ease of venting the stove into an existing chimney there. However this chimney often has no unused flues. Venting a stove into a flue that is used by any other appliance can lead to safety problems, as discussed earlier in this chapter.

Stoves in basements are sometimes partially enclosed in a sheet-metal jacket to collect most of the heat and guide it upwards through a large register in the ceiling. This makes the stove into a hot air furnace, and makes heat transfer to the upstairs more efficient. The major safety consideration is overheating of wood close to the jacket, duct, and register. Most codes do not permit direct contact between wood and anything whose temperature may regularly exceed 150° F. But temperatures of the heated air, the register, and sheet metal can easily far exceed this temperature. Thus clearances should be provided as described in Chapter 5 in the discussion of wood-fired hot air furnaces.

The fact that the stovepipe is usually partially inside the jacket can also lead to difficulties. It is usually not very tight, and leaking smoke may get into the house. It is also not readily visible for inspection, which is inconsistent with most building codes. In general, I recommend against such homemade "stove" furnaces.

CHAPTER 3

Operation and Maintenance

After improper installations, the most common cause of fires related to wood heating is negligent operation and maintenance.

CHIMNEY FIRES

Chimney fires are the burning of creosote and soot deposits inside a chimney or stovepipe connector. Chimney fires usually are started by hotter-than-normal fires in the wood heater. The heat and flames extend up into the stovepipe or the chimney and ignite the combustible deposits there.

Chimney fires are not difficult to detect. They usually involve:

1. Flames and sparks shooting out the top of the chimney.

2. A roaring sound, sometimes described as resembling a jet aircraft on takeoff, or a distant train.

3. The stovepipe glowing red hot.

4. Flames visible through small holes or cracks in the stovepipe, extending further up the pipe than normal.

5. Vibration or throbbing of the stovepipe.

Chimney fires are dangerous (Figure 3–1). Interior temperatures can exceed 2000° F. This usually causes much higher than normal temperatures of the stovepipe and on the chimney's exterior surfaces. Thus ignition of nearby or touching combustible material is more likely during a chimney fire. Proper clearances are critical during such a fire.

If the chimney has leaks, as is often the case in old unlined masonry chimneys, flames may shoot out and set fire to the house. Normally the pressure inside an operating chimney is negative relative to ambient atmospheric pressure, and air is sucked in through the leaks. Sometimes, and especially during chimney fires, positive pressures inside the chimney will push flames out any cracks. This is most likely in the upper portions of the chimney.

Stovepipe joints not fastened by sheet metal screws may vibrate apart. The roof or surrounding shrubbery can be ignited by sparks and burning creosote chunks roaring out the top of the cracks. This is most likely in the upper portions of the chimney.

Chimney fires can damage chimneys even if they do not start house fires (Figure 3–2). Masonry chimneys tend to develop small hairline cracks both in their liners and their masonry walls. Such cracks are common and do not seriously affect chimney performance although they probably decrease the chimney's ultimate service life. Intense chimney fires have caused cracks as much

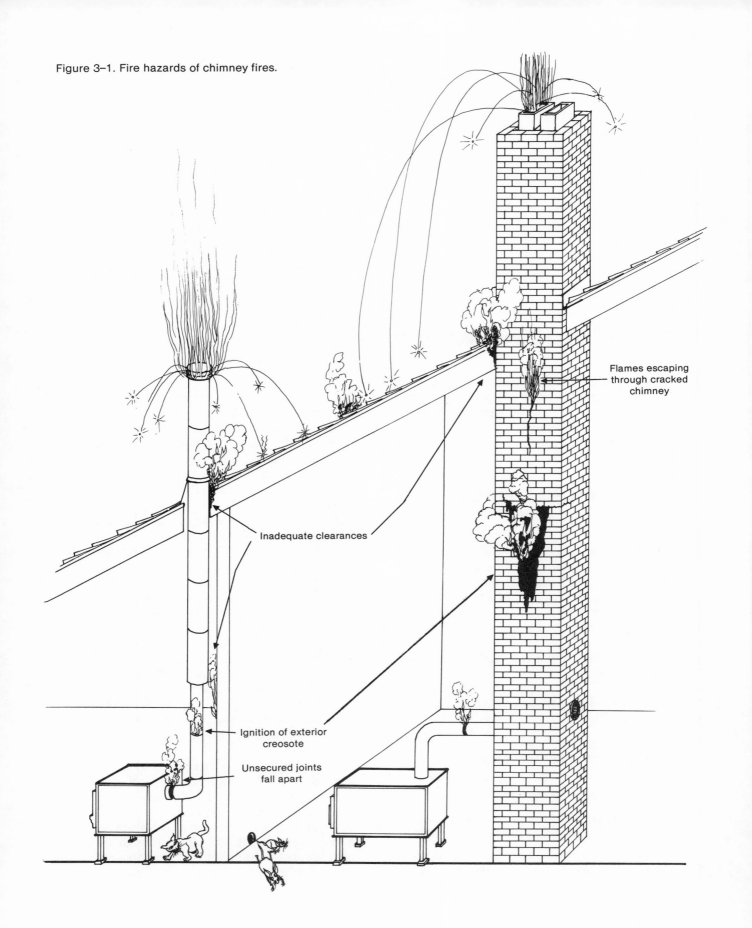

Figure 3–1. Fire hazards of chimney fires.

Flames escaping through cracked chimney

Inadequate clearances

Ignition of exterior creosote

Unsecured joints fall apart

as ¼ inch wide and as much as 10 feet long in masonry chimneys (Figure 3–3), necessitating rebuilding the chimney.

These wide cracks probably are caused in part by improper construction of the chimney—filling the space between the liner and the masonry with mortar. An airspace both keeps the masonry cooler and permits the liner to expand without stressing the masonry.

Some types of factory-built metal chimneys can also be damaged by excessive heat. Intense chimney fires can distort the inner stainless steel liner. In extreme cases crimped seams can come apart, but more usually the warping allows settling of the loose-fill insulation in some types of prefabricated chimneys. This leaves gaps in the insulation, usually at the top of each damaged section, resulting in higher-than-planned temperatures on the chimney's exterior. Damaged sections should be replaced.

Emergency Action to Suppress Chimney Fires

If you have a chimney fire, I recommend the following immediate actions, in this order:

Step 1. Shut any doors and air inlet dampers on the appliance. This should take no longer than a few seconds. If the appliance is an open fireplace without doors, go on to *Step 2.*

Step 2. Alert everyone in the house.

Step 3. Call the Fire Department.

The intent in *Step 1* is to deprive the chimney of air, for without air, the creosote cannot burn. In an ideal airtight installation of an airtight stove, closing all air inlets will suffocate the fire. If the flue is shared by another appliance, cutting off the air supply becomes harder if not impossible. It also can be dangerous, for one must be very careful that an oil or gas appliance does not come on while its chimney connector is detached or blocked.

If you have a chimney fire extinguishing flare (*not* the same as a roadside emergency flare), use it as instructed. See Appendix 5 for sources. These extinguishers emit voluminous quantities of smoke that apparently suffocates the chimney fire by depriving it of oxygen. Many fire departments use these extinguishers. More than one extinguisher may be necessary, and they should be used carefully to avoid breathing the possibly unhealthy extinguisher smoke.

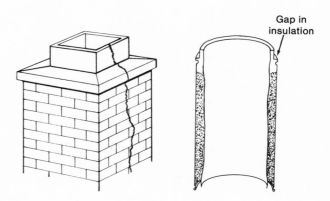

Figure 3–2. Damage to chimneys that can result from very intense chimney fires.

Figure 3–3. Cracks in a masonry chimney of the sort that can be caused by chimney fires. *Not* filling the space between the liner and the masonry with mortar reduces the chance of this happening.

Dry-type household fire extinguishers can be of some help if one can get the extinguishing substance into the chimney. However most such extinguishers run out after only 10 to 20 seconds of use. Although the chimney fire may be suppressed while the extinguisher is on, the chimney interior may still be so hot that the fire reignites.

A more down-home remedy often suggested is emptying a large box of baking soda on the stove fire This will certainly slow down the fire in the appliance. Sodium bicarbonate is a traditional fire-extinguishing chemical, but its effectiveness at suppressing fires in chimneys when it is dumped in the appliance is not well established. It certainly can't hurt.

Water should be used very sparingly, because sudden cooling of the stove, fireplace, or chimney can cause serious cracking. A large quantity of steam entering the house is also dangerous. Tossing small amounts of water, about a half-cup at a time, on the fire can generate enough steam to help suffocate the chimney fire. Putting out the fire in the wood heater is not the primary objective in controlling chimney fires. Once started, they can supply their own heat to keep going.

Depriving a *fireplace* of air is difficult. If it is fitted with doors, closing the doors and air inlets can help, but leakage through cracks is sometimes substantial. If the unit does not have doors, a large sheet of metal or asbestos board, or even a *wet* blanket can be placed over the fireplace opening. The suction from the chimney may be so great that it may take some effort to keep a cover such as a wet blanket in place.

If it seems urgent to suppress a fireplace chimney fire and all other measures have been inadequate or are not available, and if the fireplace has a damper, you can first slow down the fire in the fireplace with water (a half-cup at a time) or a large amount of sand and then gradually shut the flue damper. The result will be a very smoky and steamy house, but this is better than a burned-down house.

Keep a close watch on the house structure close to the chimney for high temperatures or smoldering wood. Water is very effective at putting out smoldering wood and keeping it cool near a hot chimney. If the wood is not yet smoldering or burning, aluminum foil placed between the wood and the chimney can provide some temporary protection. During normal operation, chimney exteriors tend to be hottest just above where the flue gases enter. During chimney fires, the hottest location depends on where the chimney fire is located; often the upper portions of the chimney are the hottest.

Avoiding Chimney Fires

There are two ways to avoid chimney fires, both of which should be followed. Do not let creosote build up to the point where a big chimney fire is possible. Do not have fires in the wood heater that may ignite chimney fires. These are hot fires, such as when burning household trash, cardboard, Christmas tree limbs, or even ordinary fuel wood, but with a full load on a hot bed of coals and with the air inlet wide open. Very hot fires are something not to be proud of but to fear. If you come close to being able to read a book by the glow of your stove, you are living dangerously, and gradually burning out your stove.

Keeping chimneys and stovepipes clean is the best insurance against chimney fires. It also improves the energy efficiency of the installation since creosote is an insulator. Cleaning as often as necessary is the key. Although there do not seem to have been many quantitative tests in this area, it is generally agreed that any creosote buildup of ¼-inch or more should be cleaned out.

Inspection for Creosote

The only certain way to know when or how often cleaning is necessary is through inspection. For stovepipes this usually requires the disassembly of a joint or two. If T's have been used in place of elbows, inspection may only require removal of caps.

Although direct inspection is preferable, the clang/thud test for stovepipe can be useful. When tapped on the outside, stovepipe with significant creosote buildup makes a duller sound (a thud) than does clean stovepipe (which clangs). Before relying on this method you should gain experience by combining it with direct inspection.

Access for inspecting the chimney itself varies. Some metal chimneys have cleanout caps in a T. In any chimney, access should be possible through the breaching by disconnecting the stovepipe. Most masonry chimneys have cleanout doors. A small mirror can be useful to look up a chimney (Figure 2–24). Some chimney tops are easily accessible. In Sweden, permanent and safe access to all chimney tops is required by law (Figure 3–4).

Chimney Cleaning

If you clean your own chimney and stovepipe, I recommend purchasing the equipment professionals use (Figure 3–5). Wire brushes are available in enough sizes and shapes to be a snug fit inside any common flue. If a chimney is cleaned from above, a weight hanging on the brush and a rope are needed to move the brush up and down the chimney (Figure 3–6). An alternative which

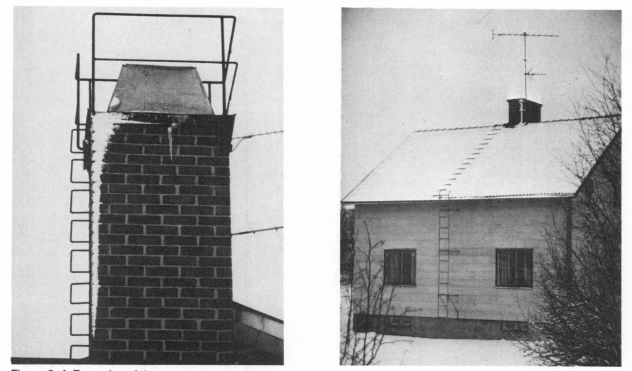

Figure 3–4. Examples of the required access to chimneys for inspection and cleaning in Sweden.

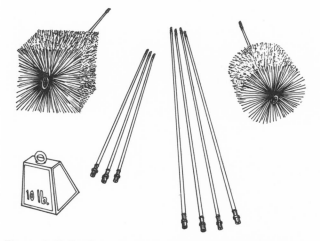

Figure 3–5. Equipment used by professional chimney sweeps.

also works when cleaning chimneys from below, is lightweight fiberglass screw-together rod sections. This equipment will make the job quicker, easier, and more thorough than most homemade methods such as using chains, bushes, or dropping a live duck down the chimney once a week. In addition, the wire brushes are less likely to damage chimney liners.

While cleaning a chimney from the roof, leave the appliance connected and all openings closed off. Put a cover over the fireplace opening. This keeps the creosote and dust contained and out of the house. After brushing, the loose, fallen creosote can be shoveled out of the cleanout area, stove, or fireplace.

When cleaning a chimney from below, most professionals use a very powerful vacuum cleaner of special design to keep the house clean. It is so powerful that with its intake nozzle just lying on the floor at a fireplace opening, one does not even need to use any shields or shrouds to keep dust out of the house. Home vacuum cleaners cannot be used this way. One way to contain the falling creosote when cleaning from below is with a rag stuffed into the bottom of the chimney or held across the chimney bottom (Figure 3–6).

If when cleaning a chimney from below there are sharp bends to follow so that the rods do not work, a plumber's snake can sometimes be used in place of the rods.

Any stiff wire brush can be used to clean stovepipe. This job is best done outdoors. Some frustration can be avoided if the order and orientation of each section is marked before taking the

75

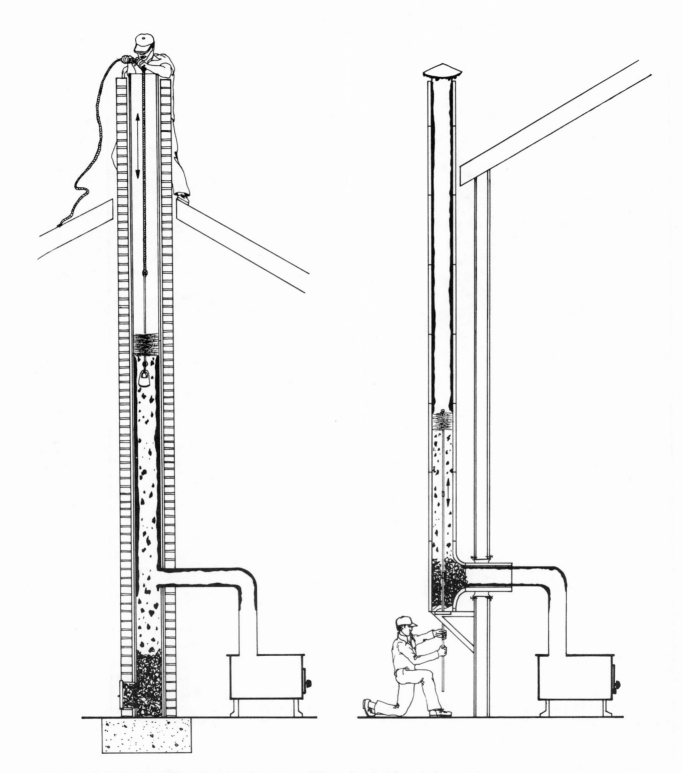

Figure 3–6. Typical chimney cleaning operations. When cleaning from below, a rag with a hole in it taped or tied across the bottom can keep the loose creosote from raining down on the sweeper. The man on the right is using screw-together rod sections.

pipe apart. This can be especially useful for realigning screw holes.

If inspection of a chimney reveals a tar-like deposit, cleaning is not worth the effort. Brushes will be fouled by the tar, and there is no safe solvent to help clean this kind of deposit. Such deposits are usually not thick enough to require cleaning. A hot fire will dry, crack, and loosen these deposits so that cleaning is easier or even unnecessary, as explained below.

Occasionally, a very hard slag-like deposit is encountered which is very difficult to clean out. A hammer and chisel, or a long iron pipe could do it, but with substantial risk of damaging the chimney. In some cases it may be best to leave these deposits alone if they are not too extensive.

Chimney caps often need cleaning more frequently than the chimney itself. The cold surfaces of most caps can accumulate creosote more quickly than the chimney surfaces.

Creosote Burnouts

Intentional small "chimney fires" once a day or once a week can help keep a chimney clean but I do not generally recommend this method. The procedure is to have a very hot fire for 5 to 30 minutes. This can usually be achieved by leaving the air inlet wide open with a full load of normal wood. The shorter times are adequate for factory-built chimneys. The longer times may be necessary with masonry chimneys.

If a real chimney or stovepipe fire is ignited, it can burn out much of the creosote. However, in practice, it is often not a chimney fire that cleans the system. In chimneys with only a light deposit of tar-like material, the heat from the hot fire dries (pyrolizes) the deposit. The creosote shrinks, and curls into flakes. These flakes then either fall to the bottom of the chimney (Figure 3–7) or are carried up and out of the top.

Regardless of the actual mechanism, extra hot fires in wood heaters are always somewhat hazardous, mostly due to the chance of real chimney fires. Small chimney fires are not as dangerous as large ones, and any chimney fire is less dangerous when it is intentional so that someone is watching it. However, a small relatively innocuous chimney fire is *assured* only if the creosote accumulation is not large, and this requires a complete and careful stovepipe and chimney inspection.

These hot-fire chimney cleanouts can work. The key to using this method safely is to be sure it *is* working every time, and not to build danger-

Figure 3–7. Creosote flakes found at the bottom of an insulated metal chimney that is "cleaned" daily with a moderately hot fire in the stove. The hot fire in this case did not burn the creosote out but caused it to fall out.

ously hot fires in the cleaning effort. Thus I would use only normal wood fuel for the hot fire, and I would inspect the chimney carefully and frequently, particularly at first, to be sure the method is working. Most chimneys treated this way will accumulate considerable loose creosote flakes in the cleanout area.

Chemical Chimney Cleaners

Chemical chimney cleaners such as *Chimney Sweep, Red Devil, Kathiteh, Safe-T-Flue*, or just rock salt have not, to the best of my knowledge, been carefully tested for their effectiveness against wood creosote buildup. My impressions are:

1. If a chimney has a buildup of ¼-inch or more, the chemicals will not do the job. Elbow grease is the only effective method.

2. If the chimney is clean when you begin using the chemicals, regular use as directed may slow down but not stop creosote accumulation.

3. The chemicals containing salt are corrosive to all types of chimneys—masonry, steel, galvanized steel, and stainless steel. Unfortunately, the seriousness of the corrosion problem does not seem to be known. If useful chimney lifetimes are reduced by 1 percent, that would not be serious. But if the life expectancy is reduced by 50 or 80 percent, I would not want to use the chemicals. And since the answer to this question seems not to be known, I do not recommend taking the risk.

How Often
is Cleaning Necessary?

Whenever there is a buildup of ¼-inch or more anywhere in the system. But there are no both simple and truthful answers, such as "once a year" or "every 3 cords." There are cases of chimney fires occurring after only one week of operation of a new stove vented to a new chimney. In some cases, cleaning is needed only every few years. In late 18th century Philadelphia, monthly cleaning was the rule. In much of northern Europe, two to four cleanings a year is standard today. Part of this variation in cleaning frequency is due to differing standards of how dirty is dangerously dirty. But the most important factor is the particular heating system and its operation. A discussion of the fundamentals of creosote will help to make clear why such differences in needed cleaning times occur.

CREOSOTE FUNDAMENTALS

Wood smoke contains these three ingredients that may end up in the chimney as a combustible deposit:

1. *Very small tar droplets* (tar fog), a major ingredient of smoke from smoldering wood.

2. *Invisible vapors.* They can come from both smoldering wood and flames wherein combustion is incomplete. Both tar droplets and vapors can condense or plate out onto surfaces such as stovepipe or chimney interiors and even inside stoves. The color is usually black or dark brown. Depending on the mixture of the tar and vapors and how much water vapor is also condensing out, the physical form of the condensate may be very fluid, like water, or tacky, like tar.

3. *Soot.* It is formed only in yellow-orange flames. When deposited, it is soft, velvety, and black. The yellow-orange color of flames does not itself imply that soot deposits will result. The color comes from glowing "yellow-hot" soot particles inside the flame, but it may (and frequently does) happen that these particles are completely burned when they reach the outer limits of the flames. As in any combustion process, adequate air and high temperatures help complete the combustion.

The term *creosote* is often used to mean slightly different things. It can refer to whatever accumulates in a chimney due to an attached wood burner, such as tar, liquids and soot, or to the tar and liquid components only, or to the liquids only, which are closely related to the fence post preservative, or to one particular chemical found in the liquid. For simplicity, I use it to refer to everything in the chimney that might burn or need cleaning, but excluding such things as birds' nests.

Once deposited inside the chimney, the creosote usually changes its physical form due to subsequent heating (Figure 3–8). The tar "dries" (pyrolizes), losing its tackiness. The deposits can become shiny curled flakes or have a bubbly appearance. As more deposits are added and baked, the accumulating mass can take on many forms—flaky granular and easily brushed, very hard and slag-like, or, if soot predominates, soft and dusty. Ash may be mixed with the creosote.

Figure 3–8. Some creosote forms.

a. Creosote dripping down on the outside of stovepipe.

d. Bubbly creosote. If, when a creosote tar deposit is pyrolized upon heating, the gases generated get trapped under the still-flexible tar, the surface bubbles up.

Effects of Chimney Type and Location

The smokiest fires and the coolest chimney temperatures produce the greatest amount of creosote accumulation (Figure 3–9). Chimney temperatures are determined mostly by chimney type and location, and by wood heater type and use.

Published information on the effect of chimney type on creosote is apparently not available. But the experience of some chimney sweeps, laboratory personnel, dealers, and users generally points to the following conclusions. The best chimney type for minimizing creosote accumulation seems to be factory-built, *insulated double-wall metal*—such as *Metalbestos*, *Pro-Jet*, and *Selkirk* brand all-fuel chimneys. This construction is apparently the best of those commonly available at keeping the flue gas heat inside the chimney.

Triple-wall metal chimneys are difficult to classify. In the thermosyphon or *air-cooled* type, intentional air flow in the two outer passageways succeeds in keeping the chimney exterior cool, but seems also to cool the flue passage, resulting in high potential for creosote accumulation. This type of chimney is rarely used with wood stoves; it is usually supplied only as part of certain metal fireplaces.

Air-insulated chimneys have less air circulation because there is at least partial blockage of the passageways between chimney sections. Thus the air-insulated metal chimneys tend to be better at minimizing creosote than the air-cooled types. However, I have seen no triple-wall chimneys that truly isolate the air inside each passageway within each chimney section. Some air leakage to or from the outside also likely occurs at each junction. The

b. Pool of creosote on floor under stovepipe. When it is this fluid, creosote contains much water.

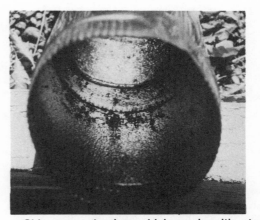

c. Shiny creosote glaze which may be either tacky (tar) or dry (mostly carbon) depending on how hot it has been.

e. Creosote flakes, caused by the drying and shrinking of creosote glaze (c).

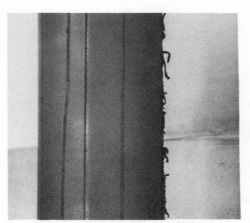

f. Creosote curls, a dried form of creosote drips (a).

79

distinction between air-cooled and air-insulated chimneys is further blurred by the fact that some air-cooled chimneys do not thermosyphon very effectively, probably due to leakage between passageways and to the outside. Clearly some comparative laboratory testing of all major brands of metal chimneys would be useful.

It is not safe to modify any factory-built chimney in an effort to increase its ability to hold in the flue gas heat. For instance, a fully sealed air-insulated chimney might get hotter on its exterior and thus require more installation clearance. In fact this may be one reason there are no truly air-insulated metal chimneys (chimneys with no convection between sections and no escaping of the heated air).

Masonry chimneys, despite their thickness, are not very good at keeping flue gases warm. Their high mass is in fact a disadvantage here—each time a fire is started when the chimney is cold, the chimney stays cold longer than any other type. Thus some creosote buildup is almost inevitable at the start of each fire. But even when warm, their heat loss rate is high.

Bare single-wall metal pipe, such as stovepipe or culvert pipe, is the worst, having the highest heat loss rate and the most creosote accumulation. In my wood heater testing laboratory, I have installed alternating stovepipe and insulated chimney sections in the venting system for scientific purposes. The difference in creosote accumulations is very clearly visible. (This installation is not legal and not recommended in homes.)

Oversized chimneys tend to accumulate much creosote. This is often a problem with stove installations in or in front of masonry fireplaces. The excessive size of a fireplace flue results in the wood stove smoke rising only very slowly in the chimney. This gives the smoke more time to cool, and the large surface area of the flue also con-

g. Soot, mixed with fly ash. Soot is very soft; a swipe with a finger rubs it off, as shown in the photograph.

h. Creosote slag. Accumulated creosote on cold surfaces can build up into a very hard, slag-like deposit, in this case on a water heating coil wound inside stovepipe.

i. Creosote deposits resembling fungus growth on dead tree. The aerodynamics of smoke flow may have had some effect here.

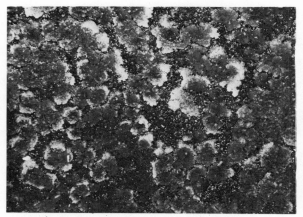

j. Creosote on the inside of a wood boiler.

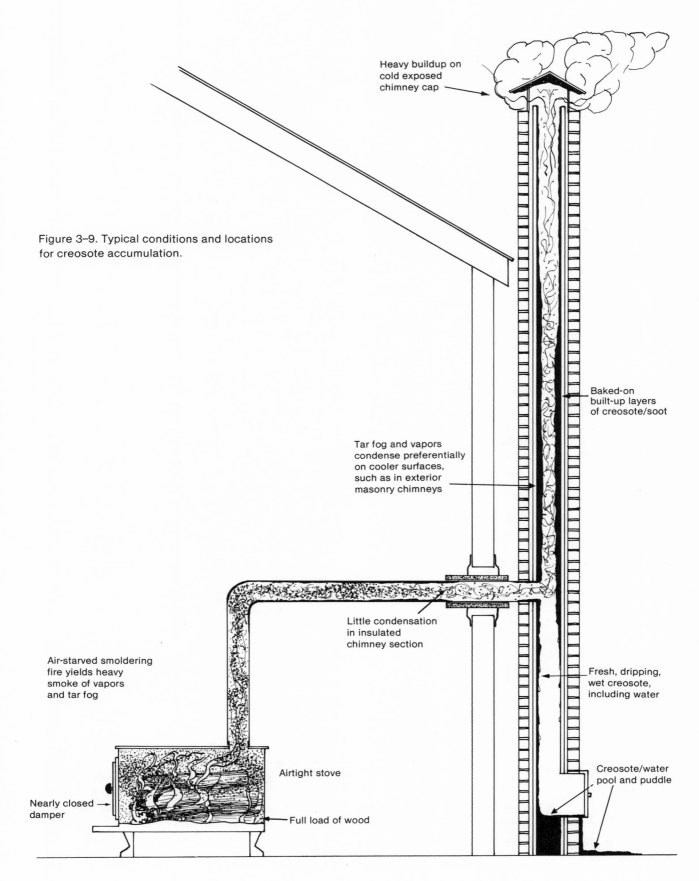

Heavy buildup on
cold exposed
chimney cap

Figure 3–9. Typical conditions and locations
for creosote accumulation.

Baked-on
built-up layers
of creosote/soot

Tar fog and vapors
condense preferentially
on cooler surfaces,
such as in exterior
masonry chimneys

Little condensation
in insulated
chimney section

Air-starved smoldering
fire yields heavy
smoke of vapors
and tar fog

Fresh, dripping,
wet creosote,
including water

Airtight stove

Nearly closed
damper

Creosote/water
pool and puddle

Full load of wood

81

tributes to the cooling. If the fireplace chimney is an exterior chimney, the problem can be very severe.

Installing a stove in or in front of a metal fireplace with an air-cooled chimney can result in the two worst features of chimneys for creosote—being oversized and cold. I recommend against such installations. Inadequate draft and flow reversal are also possible problems.

The better chimney *location* is running up through the interior of a house; the worse is on the outside of a house (Figure 3–10). The cooler temperature air around an exterior chimney causes cooler temperatures inside the chimney; this results in more creosote. Of course, even interior chimneys have exterior portions at the top and in some cases heavy creosote accumulates only above the roof line, indicating the importance of surrounding temperatures. In addition to decreasing creosote accumulation, interior chimneys increase draft, increase system energy efficiency, and decrease susceptibility to flow reversal.

My personal preference, not considering cost or convenience of installation in an existing house, is an interior masonry chimney with all its walls exposed to the living spaces. By trying to avoid smoldering fires I manage to avoid much creosote buildup, and the exposed masonry contributes considerable heat. However I have installed some prefabricated metal chimneys in my homes because of the ease of installation.

Some manufacturers claim their chimney caps are effective at reducing creosote accumulation in chimneys. Because I have seen no careful testing

and because the manufacturers' explanations are not convincing to me, I do not recommend investing in such caps for their creosote-inhibiting effects. Most caps *are* effective at keeping out rain and minimizing wind effects on chimney draft. Use of conical diameter-reducing caps on top of oversized chimney flues may improve draft a little.

Effects of Equipment Selection

Stoves and accessories that extract more than the average amount of heat from the fire and flue gases produce cooler smoke and more creosote accumulation (Figures 3–11 and 3–12). Examples are:

1. Stoves with an extra chamber for better heat transfer because of the extra surface area and residence time for the smoke, such as the double barrel stoves, the Sevca, the Jøtul 606, the Morsø 2BO, and the Lange 6303 and the Ulefos 172.

2. Heat-extracting accessories, also called heat savers and heat robbers.

3. Installations with extra long lengths of stovepipe between the stove and the chimney.

All these features increase the energy efficiency of the system; thus there is often a conflict between maximizing energy efficiency by extracting more heat, and minimizing creosote.

Oversized equipment tends to result in more creosote. A stove designed for 60,000 Btu per hour that is used to heat a single room needing only 10,000 Btu per hour will be operating at the low end of its power range where the tendency is towards smoldering, smoky fires, and creosote.

The surest and always available way to decrease creosote accumulation is to decrease the smokiness of the fire. If combustion is complete, the flue gases consist only of water vapor, carbon dioxide and air—nothing to condense out to form creosote. Improving combustion efficiency can also improve overall energy efficiency, but only if it is done without decreasing heat transfer efficiency.

How does one obtain complete combustion? The single most important condition is to have plenty of air available to the fire. Fireplaces have very high air-to-fuel ratios and generate much less creosote, especially on a per-pound-of-wood basis, than do stoves. Stoves whose combustion air cannot be severely limited, such as the traditional Franklin stove which admits substantial air

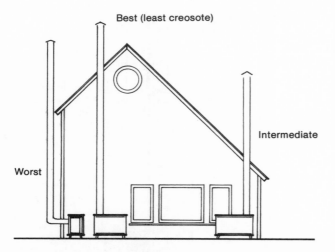

Figure 3–10. The effect of chimney location on creosote accumulation.

even with the doors and air inlets shut, tend to generate less smoke than do airtight stoves.

Airtight stoves have a justified reputation for generating the most creosote. The worst thing to do with these stoves is to put a full load of fuel on a bed of hot coals and then to turn the air inlets or thermostat down low. This is how to achieve the maximum burn duration, such as overnight, but it also results in the most creosote. The hot wood generates smoke, but without ample air the smoke cannot burn.

Most special design features intended to reduce creosote by burning the smoke more completely are not very effective. In my testing of stoves I have studied secondary air systems in three stoves, firebrick liners in two stoves, and baffles (of the Jøtul 118 type) in two stoves. The stoves were chosen to be typical of products on the market today. The tests consisted of comparing combustion efficiencies with and without the liner present, with and without the baffle, and with the secondary air inlet open and closed.

I found no *significant* average effect on combustion efficiency due to these features on the stoves tested.

Secondary air systems add air to the smoke to help it burn. I found they did not work when needed the most—during smoldering, full-load, limited-air burns. To get smoke or anything else to burn requires both adequate air *and* sufficiently high temperatures. But the very smoky burns have the coolest smoke temperatures, and adding extra air tends to be ineffective.

Liners and baffles *can* be useful for other purposes. Liners can increase the durability of the

Figure 3–12. A high heat-transfer-efficiency stove in the National Historical Museum in Copenhagen. Note the chair for scale.

Figure 3–11. Examples of equipment with high creosote potential by virtue of high heat transfer efficiency.

83

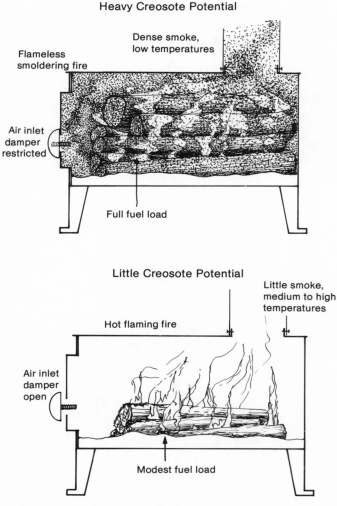

Heavy Creosote Potential

Flameless smoldering fire

Dense smoke, low temperatures

Air inlet damper restricted

Full fuel load

Little Creosote Potential

Little smoke, medium to high temperatures

Hot flaming fire

Air inlet damper open

Modest fuel load

Figure 3–13. Creosote potential as affected by operating procedure.

Figure 3–14. Heavy creosote deposits inside a stove right next to the fire. This is unusual, requiring exclusively very cool, smoldering fires. This is most likely to happen when the stove or other wood heater is oversized for the heating job.

appliance, liners and baffle can improve the steadiness of the heat output, and baffles can increase heat transfer efficiency. Thus they can be desirable features even if they are not very significant at reducing smoke output from wood heaters.

There may be wood heaters with effective creosote-reducing features I have not tested. And there continues to be research into really effective smoke "consuming" devices. However at the present I feel one must rely on careful *operation* of wood heaters to minimize smoke output.

Operation

Most airtight stoves can be operated to yield as clean a burn as any other wood burner. What is required is, of course, different for different stoves and fuel. Typically, smoke-free clean burning requires small fuel loads—two to four logs at a time, or ¼ to ½ of a full load—and leaving the air inlet relatively wide open, especially during the first 10 to 30 minutes after each loading, when most of the smoke-generating pyrolysis reactions are occurring (Figures 3–13 and 3–14). Towards the end of each burning cycle the air inlet can be turned down substantially without fear of smoke generation; charcoal (when pure carbon) cannot generate creosote-producing smoke.

In my own home, I try to use the appearance of the flue gases coming out the top of the chimney as a guide (Figure 3–15). If I see much smoke I open the air inlet further, or remember to use less fuel next time. If the flue gases are clear, I cut back on the combustion air. In this mode of operation, the rate of heat output is essentially determined by the size of the fuel load. In mild weather, small fuel loads must be used; otherwise too much heat will be given off when enough air is admitted to achieve clean combustion. In cold weather, substantial loads, perhaps even full loads can be used. Because many people tend to use large fuel loads all the time, some manufacturers of large central wood-heating boilers recommend wood not be used at all in spring and fall but that only the backup gas, oil or electric system be used. This is one way to decrease creosote, but another, of course, is to use smaller and more frequent loadings in the spring and fall.

Constant checking and fiddling is, of course, impractical and unnecessary. The real and necessary sacrifice is more frequent refueling of the wood heater—perhaps every 1 to 3 hours instead of every 3 to 9 hours. If the long duration, large load, unattended burn is important, more frequent chimney cleaning will usually be required. On the other hand, if you are at home most of the time and do not mind tending the stove every 1 to 3 hours, using smaller, more frequent loadings will usually decrease creosote accumulation significantly.

Testing with a Jøtul 602 stove, I've found that overall energy efficiency does not seem to be affected much by this choice. About the same total amount of wood is consumed to produce the same

total amount of heat in either case—large infrequent fuel loading, or more frequent but smaller loadings.

Monitoring Smoke Temperature

Since the temperature in a chimney is so closely linked to creosote condensation, it is often thought that measuring the smoke temperature and

A   B   C

D

Figure 3–15. Use of chimney smoke to gauge combustion efficiency. With no visible smoke, combustion is relatively complete and there can be little creosote accumulation. Steam is not smoke. The photographs illustrate the difference. A. Heavy smoke. Smoke consists of tar droplets and soot particles which cannot simply disappear. Thus downwind from a smoking chimney a smoke haze persists. B. No smoke, indicating relatively complete combustion. (There *is* a large fire in the attached wood stove.) C. Steam, but no smoke. Pure steam also indicates complete combustion. Combustion of wood or coal always generates water vapor, even if the fuel is very dry. This vapor is invisible as it rises inside the chimney, but when it encounters the cold outside air, it can condense into droplets that are then visible as steam. As the steam spreads out, the droplets evaporate and the water becomes invisible again. Thus downwind (or up-plume) from the chimney the air is perfectly clear. In some cases a second identifying feature of steam is a clear, steam-free, space of a foot or so just above the chimney. The invisible water vapor there has not yet cooled enough to condense into droplets. D. Often the plume from a wood heater chimney contains both real smoke and steam. After the steam droplets have evaporated the plume is less dense.

keeping it above a critical level would prevent creosote accumulation. Unfortunately the situation is more complex.

Typically the only convenient place to measure the smoke temperature is in the stovepipe connector to the chimney. (A stem thermometer whose heat-sensitive element is inserted into the pipe is better than a surface thermometer.) But the temperature in the stovepipe is not the same as the temperature in the chimney. The smoke cools as it moves along. The cooling can be very significant in uninsulated chimneys, in very tall chimneys, in exterior chimneys, and in oversize chimneys.

Even if one could measure the temperature of the smoke throughout the chimney, this would not be directly useful. More relevant is the temperature of the interior wall surface of the chimney, for that is where the creosote actually condenses.

Finally, there is no critical temperature, wherever measured, such that no creosote accumulates for higher temperatures and creosote accumulation is heavy for lower temperatures. There is probably a wide, critical-temperature *range*, and the range itself probably depends on smoke composition, due to wood type and moisture content, and smoke density.

Despite all these uncertainties, I would guess that with smoke temperatures (measured with a stem thermometer or thermocouple) above 250–300° F., little creosote will accumulate. The temperature measurement would have to be made at the top of the chimney to assure temperatures exceed that range throughout the chimney.

Monitoring temperatures can be fun. I have stem thermometers in all the stovepipe connectors in my home. The thermometers indicate at a glance the intensity of the fire and can warn of dangerously hot conditions. (No chimneys are designed to take flue gas temperatures above 1000° F. continuously—especially masonry chimneys; see Appendix 3.) A useful and commonly available range for a stem thermometer is 200–1000° F. Surface thermometers need not go above 500 or 600° F. (See Appendix 5 for sources for thermometers.)

Effects of Fuels on Creosote

Properties of the fuel used also affect creosote accumulation, but they are not as important or as easily controlled as the load size and air settings. Most important, *use of "seasoned hardwood" does not eliminate creosote*. In fact, wood can be *too* dry. My guess is that in closed stoves, the best

moisture content is between 15 and 25 percent, both for minimizing creosote and for maximizing overall energy efficiency.[1] When wood is drier than about 15 percent, more smoke and creosote result in systems with air-limited combustion. Dry wood burns so easily that the air supply must be more limited to avoid too much heat output, and less air almost always means more smoke. In addition, drier wood heats up more quickly in a fire because it has less moisture which requires heat to be driven off. Thus the whole fuel load tends to be undergoing pyrolysis simultaneously, and more quickly than with moister wood. This results in a larger surge of smoke, and since combustion tends to be air-limited under these circumstances, most of the smoke cannot burn for lack of oxygen. The conventional wisdom among wood burners is that green (moist) wood produces more creosote. This does not necessarily contradict my experimental results. The conventional wisdom may be based on experience with fireplaces and non-airtight stoves. My experiments were done with an airtight stove.

Does fuelwood ever get drier than 15 percent moisture content? Not often, but it can. Wood in typical fuel wood size pieces, when stored in a heated (70° F.) space, can get down below 15 percent moisture content in about a year. After two years the wood will be near its equilibrium moisture content, which depends mostly on the relative humidity of the surrounding air, and is about 6 to 10 percent. In New England I have found that wood stored outdoors and uncovered for a year or more has a moisture content of 20 to 30 percent.

Very small pieces of wood produce more smoke and creosote when burned in a closed stove than do normal size pieces. The wood heats more quickly, resulting in a larger surge of smoke, a smaller fraction of which burns in the air-starved combustion regime in most stoves. Using scrap lumber and wastes from wood-products manufacturers often results in considerable creosote, because of the small pieces and the excessive dryness of the wood.

Pine has a reputation for creating more creosote accumulation in chimneys than do dense hardwoods. There is also some evidence to the contrary. It may depend on the type of wood heater

---

1. For more details on the experimental basis for these conclusions, see J.W. Shelton, "Wood Stove Testing Methods and Some Preliminary Experimental Results," ASHRAE Trans. 84, Pt. 2, pp. 388–404 (1978). These conclusions have been confirmed by other researchers (private communication with Robert Jorstad).

used, on the moisture content, and on the species of pine. The principal disadvantage of most pines is the low energy content per cord. Since each log has less energy than, for example, that of oak of the same dimensions, more frequent loading of the wood heater will be necessary, and if wood costs the same per cord, the economics of pine are less favorable. Thus, given the choice, denser woods are favored because they require less frequent refueling, usually yield more heat per dollar spent on fuel, or, if you prepare your own fuel, require less effort for the same heat content. So regardless of the creosote potential, there are other sound reasons to select dense hardwoods over pines when there is the choice. In practice, of course, people burn what is available and all wood species can be used to keep you warm.

Creosote Traps?

Trapping or extracting some of the creosote out of the flue gases before they get to the stovepipe or chimney may be possible. However, since the temperature of the smoke and the surrounding chimney walls will decrease as the smoke moves through the venting system, some additional creosote is likely to condense out regardless of how much the creosote trap took out.

I know of no creosote traps on the market, and there may never be. However, there are unsubstantiated but plausible reports that large wood boilers (water heaters) produce less chimney creosote because so much creosote accumulates in the relatively cool inner surfaces of the boiler. Typically the water keeps them at about 200° F.

or less. The boiler surfaces need frequent cleaning, usually every 1 to 4 weeks. This job can be very quick and easy, requiring 5 or 10 minutes of scraping or brushing, and the residue all falls into the combustion chamber or ash tray.

Is creosote all that bad? Does it merit pages of instructions for avoiding it? Perhaps not. The major safety issue related to creosote is chimney fires, and frequent enough cleaning can always eliminate this hazard. Thus for those who are willing to clean their chimneys more often, creosote accumulation may not be a "problem." However another potential problem caused by incomplete combustion in wood and coal heating systems is air pollution. It is not yet known how unhealthy wood smoke may be. But smoke is clearly at least a significant visual pollutant in many communities. Thus operating solid fuel heaters for clean combustion has the added benefit of resulting in cleaner air.

Back Flashing

Care must be taken in opening the door of any wood heater in use. Burning wood can fall out, although this is rare in most stoves. Smoke and sometimes flames can emerge, but usually only very briefly.

A more unexpected and dangerous phenomenon is back flashing or puffing (Figure 3–16). Airtight wood burners, including stoves, furnaces, and boilers, are the most susceptible. The combustion rate in airtight stoves is usually air-limited as opposed to fuel-limited. If air is suddenly admitted by opening the door there can be a

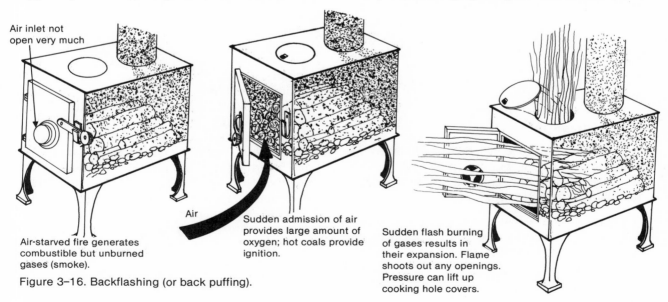

Air inlet not open very much

Air-starved fire generates combustible but unburned gases (smoke).

Air

Sudden admission of air provides large amount of oxygen; hot coals provide ignition.

Sudden flash burning of gases results in their expansion. Flame shoots out any openings. Pressure can lift up cooking hole covers.

Figure 3–16. Backflashing (or back puffing).

combustion surge of the gases which is so quick that the resulting pressure can force hot gases and flames out any available openings, such as the loading door, the air inlet, and cracks or leaky joints in both the stove and stovepipe. The phenomenon could be termed a slow, very small explosion. Back flashing is dangerous because the emerging gases may be burning, and thus can burn people nearby. To avoid back flashing and momentary leakage of smoke, the air inlet should be fully opened for half a minute before the door is opened. Then the door should be opened slowly. The operator's face should be kept well back until the door has been fully open for a moment. The danger of back flashing is one reason why children should be cautioned about operating wood stoves.

Back flashing is most likely to occur after a load of wood is added on top of a bed of hot coals and the air inlet is turned down. The stove may fill with combustible gases. If more air is suddenly admitted, such as when opening the door, or sometimes even via the air inlet, the gases may suddenly ignite, creating a substantial back flash. Back flashing can also occur spontaneously at any time. On rare occasions the pressure within the stove will lift heavy metal cooking and warming plates. Another occurrence is pulsating back flashing— small flashes occurring consecutively, sometimes separated by only a half a second, sometimes separated by as much as five seconds. Each flash consumes the available oxygen and pressurizes the stove, momentarily preventing more oxygen from getting in. The next flash will occur when enough oxygen has reentered the stove.

The phenomenon of back flashing deserves respect and perhaps some research effort. Particular kinds of stove designs may be especially susceptible, and others may be immune. I have never heard of a stove exploding due to this phenomenon, but the blowing off of a cooking plate clearly leaves a stove in a more dangerous condition.

Fire Extinguishers

It is wise in any home, whether wood-heated or not, to be prepared to put out small fires.

In case of fire, the generally recommended emergency actions, in order, are:

1. Alert all the people in the house and either get them out or near exits which cannot become unusable if the fire spreads.

2. Call the fire department.

3. If the fire is small and if you have the equipment and if you can work where a spreading fire could not block your escape, and if you have the sense to get out when the battle is being lost, then try to put the fire out.

The appropriate way to fight a fire depends on the type of fire. If the fire involves ordinary combustible materials such as wood, paper, and fabric ("Class A" fires), one needs extinguishing agents that cool and/or coat the burning material. Any water-based extinguisher is effective, and water itself is excellent. An extinguishing agent, such as carbon dioxide, that merely and temporarily excludes air is not adequate, for as soon as the extinguisher is turned off the residual heat in the burning material may reignite the fire.

"Class B" fires involve burning liquids or gases—grease, lighter fluid, or kerosene. Here water is not very effective since most flammable liquids float on water. Also, throwing water on a frying pan of burning grease may splatter the grease all over the room and create a huge amount of steam. This can be dangerous to the person trying to put out the fire. Baking soda and salt are useful on this type of fire. Carbon dioxide, dry chemical, and foam extinguishers are also effective.

"Class C" fires are any fires involving live electrical equipment where the safety of the firefighter requires that there be no electrically conducting path between the fighter and the electrical equipment. This precludes the use of water, as electrical current could flow from the equipment through the water stream and to the person. Carbon dioxide or dry chemical extinguishers are generally recommended.

New fire extinguishers are clearly marked as to fire-type for which they are designed—A, B, or C, or combinations. For most kitchen fires, a combination B and C unit is usually appropriate. Some dry chemical type extinguishers are rated for all three types of fires.

Most wood-heating related fires are Class A fires and thus require Class A extinguishers. The "old" inverting type, which is turned upside down to activate, *is* water based but is no longer recommended for use on any fire. The reasons are that these extinguishers are potentially dangerous to the operator during use, the agent is more corrosive than plain water, they are costly to service properly, they cannot be turned off once activated, and the agents are extremely good conductors of electricity, more so than plain water.

I recommend either a modern extinguisher rated for Class A fires or water. A bucket of water and a cup kept near a stove can handle many fires. Portable hand-pumped extinguishers are also good if tested periodically and kept in working order. Some more fanatical homeowners have "fire hoses" in their homes, often a high quality garden hose with a spray nozzle, that is hooked up to a water source. If the hose is long enough to reach throughout the house such a system is very effective.

Most commercial extinguishers must be recharged after every use no matter how brief, since pressure will gradually be lost. The total available duration of the blast from most extinguishers is typically only about 10 seconds. The extinguishing power can be substantial during that time, but if it fails to do the job, something else will be needed.

Ash Disposal

Every year many fires are caused by emptying ashes into cardboard boxes or paper bags. My uneducated intuition used to be that if there had been no fire in a stove or fireplace for at least a day or two, the ashes would not be hot enough to ignite paper or cardboard. Wrong! Charcoal buried in ash can stay red hot for days! Ash is a good thermal insulator and also keeps enough oxygen away so the charcoal does not burn up. This is why it is important to empty the ashes from a stove or fireplace into a noncombustible container, such as a metal container.

Additional precautions that may also be helpful in extreme cases are to keep a tight-fitting metal cover over the ash container and to keep the ash container itself off combustible floors and not touching any other combustible materials. The cover will keep any noxious fumes inside and will keep oxygen out, thus inhibiting combustion of charcoal.

Maintaining Clearances

Clearances between stoves and any combustible material must be maintained. Thirty-six inches is the legal distance for unlisted radiant stoves, not only to unprotected walls, but to furniture, drapes, logs and kindling, boxes, pillows, newspaper piles, and matches. Rugs and blankets on the floor should be kept at least 18 inches away from the stove sides.

Figure 3–17. Unsafe ash storage container and location. Hot coals in the ash burned portions of the paper bag containers. The wood siding of the house is slightly charred. (Photo courtesy of Leonard Fontaine)

Clearances from chimneys must also be maintained. Furniture, clothing, and drapes should be kept at least 2 inches away from most types of masonry and factory-built chimneys. Trees, shrubs, and vines should be kept trimmed back away from exterior chimneys.

Stovepipe Maintenance

Stovepipe does not last forever. Corrosion is accelerated if wet creosote condenses in the pipe and if household trash is burned in the wood heater. Plastics containing fluorine or chlorine emit very corrosive gases when burned. Thus periodic inspection and replacement are necessary. This is why most codes require that stovepipe be exposed and accessible and one reason why many people discourage passing a stovepipe through a wall even when it is done properly.

Chimney connectors should be kept mechanically secure, with sheet metal screws or the equivalent at every joint, including the connection to the stove. Unfastening and fastening screws, and lining up screw holes in stovepipe can be bothersome, particularly if frequent disassembly for inspection and cleaning is practiced. However, it is necessary for safety. Chimney fires can be physically as well as thermally violent events, causing unfastened joints to slip apart. Marking the stovepipe sections for their order and orientation can be useful.

Sparks

Open fireplace stoves and fireplaces are the most common sources of fire-producing sparks.

Open fires should always be watched attentively. Better yet is to avoid having a fully open fire by using a spark screen or glass doors when a view of the fire is desired. Tests in my laboratory of one fireplace screen indicated only a very small reduction in heat output due to the screen. Since it is inevitable that open wood heaters will not always be watched carefully or covered with a spark guard, adequate floor protection around such units is especially vital. It might be wise to have a metal dust pan or shovel and a poker or asbestos gloves to use to get an errant, hot ember back on the floor protector or back into the fire.

Avoiding Dangerously Hot Fires

Chimney fires and fires due to inadequate clearances are usually ignited by especially hot fires in wood heaters. Some fuels are obviously prone to making hot fires. These include cardboard, trash, Christmas present wrappings, and Christmas tree branches. Some less obvious causes of hot fires are more insidious because they occur when unexpected, for instance, when people are out of the house or asleep. Failure to fully latch the door after filling a stove with wood may seem unlikely, but it apparently has caused

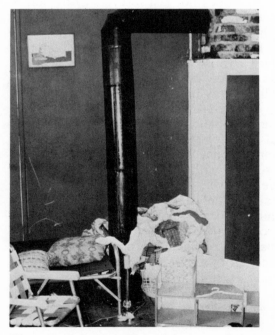

Figure 3–18. Inadequate maintenance of clearances. Clothing, quilts, and furniture must be kept a safe distance away from chimneys and stovepipe connectors. This installation is unsafe in other respects. The stovepipe connects to a wood stove one floor below. Codes prohibit passing stovepipe through a ceiling or floor. (Photo courtesy of Leonard Fontaine)

serious fires. Accidentally leaving the air inlet control wide open, or even intentionally setting the control but getting a hotter than usual fire can lead to trouble. Even properly operating thermostatic controls can be accidentally set too high or may not close down until after the fire is very hot. Non-airtight stoves sometimes get uncontrollably hot, even with the explicit air inlet shut, by using the leaked air for combustion.

Very hot fires are also very hard on the wood heater itself. Stoves can be gradually burned out, and the rate of oxidation increases dramatically with temperature. A ¼-inch steel plate held at red-orange hot temperatures for a day can disappear. If you want your stove to last a long time, never let it get close to glowing-hot temperatures. Coal should not be burned in an appliance not designed for it since coal fires tend to be hotter than wood fires.

Other Dos and Don'ts

Drying out wet or green wood on top of a stove may be tempting but the all-too-possible consequences of a fire outside of the stove are not.

Do not use lighter fluid, kerosene, or gasoline to light a fire. Use of lighter fluid outdoors is *much* less dangerous than indoors.

Do not completely shut the damper on an active fire in an airtight stove venting into a non-self-starting chimney. Flow reversal is likely if you do.

Always leave about an inch of ash in the bottom of wood-burning appliances without grates and ash drawers, or without firebrick floors. The ash protects the stove bottom and the floor beneath from overheating. Stoves with grates and ash pans or with firebrick liners may be cleaned of ash completely.

Avoid burning corrosive fuels in wood heaters, such as some plastics in household trash. Since the degree of corrosiveness of salt-water driftwood is not known, I would be cautious about burning it regularly.

Strongly warn children not to operate wood and coal heaters. Carefully instruct guests and sitters about safe operation.

Leather gloves and most kitchen gloves are useful for picking up hot coals that have fallen out of a stove. However, leather and most kitchen gloves are combustible, thus do not hold onto a hot coal hard or long, and be sure every bit of hot coal is off the glove when done.

Some clothing is extremely flammable. Particularly when wearing loose clothing, be careful when working around a fire or hot stove.

CHAPTER 4

Safety and Wood Stove Design

The overwhelmingly important factors in wood-heating safety are installation, operation, and maintenance of the system. However, there are a few characteristics of the wood-heating equipment itself that have safety aspects.

Wood-heating equipment can be dangerous in three areas.

1. *Basic design* can be so grossly negligent that the equipment cannot take normal use without critical components suddenly cracking, breaking, melting, burning, or exploding. For instance, aluminum should not be in close contact with flames, glowing charcoal, or hot flue gases since it will melt. Since cast iron can crack, cast-iron stoves should be designed so that if a casting cracks clear through, the pieces will be held in place by the rest of the stove. Since steel plate and cast iron can warp, designs must allow for warping, or at least stress, without the appliance falling apart or developing large gaps between critical parts. Any water-heating appliance or accessory must be designed to prevent it from exploding should overheating increase the pressure inside. With the possible exception of water-heating systems, such structural inadequacies are rare in practice. Use of common "tried" materials and use of standard engineering practices can almost completely eliminate this potential problem area.

2. More subtle *design and manufacturing oversights or compromises* may result in slightly but rarely catastrophically hazardous equipment. Examples are unbaffled air inlets, non-positive fuel door latches, poorly fitted spark screens, electrical components not designed for high temperatures, glass doors that break too easily, missing bolts, inadequate welds, and use of defective parts. Some defective equipment will always slip through the best of quality control programs, and few safety-related design features are necessarily always desirable, all things considered. (The very safest stove would be one that is welded shut so that it cannot be used.) Thus since no woodheating appliance is totally free from all of these problems, it is more difficult in this area to set standards. In addition, installation, operation, and maintenance of woodheating systems can have a very large impact on the actual hazardous effects of some of these aspects of wood heaters. Thus one can only point out design features to look for, but not state categorically that such features are always critical or even necessarily important.

3. A difficult area of wood heater design safety to deal with is *durability* with respect to gradual deterioration, such as burnout of fire chamber walls. Durability affects safety only indirectly, only after sufficient degradation has occurred. All

woodheating appliances, and chimneys and stovepipe connectors especially, gradually wear out and need repair or replacement. Life expectancies of equipment and parts are also very much affected by how the system is operated. Thus there are not specific equipment design features that relate directly to safety. Presumptions of both how the equipment will be used and an arbitrary desired life expectancy are necessary before one can call a thin-walled stove "unsafe." And yet, particularly for the careless user who does not notice or respond to the deterioration of his equipment, a more durable stove is safer because dangerous situations will occur less frequently.

Wall Thickness

Wall thickness in stoves can vary from ¼-inch or more in the heaviest steel plate or cast-iron stoves down to 0.020 in. in $19.95 "disposable stoves" (Figure 4–1). The thinness of a stove's walls does not make the stove more dangerous to use as long as the walls are mechanically sound. Thin-walled stoves can have slightly hotter sides than a similarly designed but thick-walled stove[1] but since installation clearances and wall protectors are either designed for the worst-case stove or are specifically tailored to each stove, this should not result in an unsafe situation. However, thin-walled stoves will not last as long as thicker-walled models, and the appliance is dangerous to use when the walls start to deteriorate.

Thus thin-walled stoves are only less safe if left unchecked, and not repaired or replaced when needed. The same attention is required to keep stovepipe connectors safe. And since maintenance is always less than ideal, thick-walled stoves are statistically safer.

Thermostatic Controls

Thermostatic controls (Figure 4–2) on stoves are intended to insure automatically a steady room temperature as long as there is fuel to be burned. A bimetallic coil senses a combination of room air temperature and stove surface temperature, and in response, opens or closes the air inlet damper, which increases or decreases the fire intensity.

Safety arguments can be made both for and

---

1. Specifically, the emitted radiation may be more intense. Lack of as much lateral thermal conductance in thin walls results in "hot spots." Thus if average surface temperatures were the same, thin-walled stoves would emit more radiation due to the absolute-temperature-to-the-fourth-power dependence for radiated energy.

Figure 4–1. A thin-walled stove, 18 inches high, purchased for $19.95 in a New England general store in 1978. The wall thickness is about 0.015 inches, the same as much available stovepipe, and less than is recommended for stovepipe by NFPA and building codes. There is an inner sidewall liner of 0.011 in. steel. The stove weighs 8½ pounds. Such stoves could almost be described as "disposable."

against thermostatically controlled combustion air. If the temperature of the control's coil is directly and principally influenced by the stove's surface temperature, the control will limit stove and flue gas temperatures. This is good for safety, but not necessarily for controlling room temperature. If the control principally senses room temperature, its response to an overheated stove will be slower and less effective from a safety standpoint since the control will hold the air inlet open until the room warms up to the thermostat setting.

To satisfy both the safety and comfort objectives requires two sensors, one exposed to room temperature and one to flue gas or stovepipe or stove surface temperatures, and they should be designed so that the latter overrides the former. Since too fast a time response for the overheat sensor would be inappropriate, some thermal mass (or electronic time averaging or delay) ought to be incorporated. Whether such complexity would yield significant returns in safety is not guaranteed but is worth investigating.

A potential disadvantage of thermostatic controls is possible malfunction leading to maximum air being supplied to the fire. It is alleged that some controls can become stuck open, and that some can be held open by rushing incoming air, such as during a chimney fire. It is vital that under any conceivable malfunction or dangerous circumstance the unit should shut, not open. Underwriters Laboratories has incorporated this principle in its proposed safety standards for stoves. (This is just the opposite of the energy-saving chimney damper controls for gas and oil appliances, where, to assure venting, any imaginable mechanical failure must result in the damper being open.)

A thermostatic control could also cause difficulties when operating properly. If an outside door or window is left open, or if a window is broken, and if no one notices or acts, the constant cooling of the house may signal the thermostat to keep the air inlet wide open, resulting in dangerously hot conditions around the stove and its chimney.

On balance, I find no convincing case for or against the safety aspects of thermostatically controlled combustion air systems on wood heaters.

Door Latches

A number of house fires have apparently resulted from stove doors not being latched. The consequences can be burning wood falling out or a dangerously hot fire due to the large air supply. The direct cause of such fires is of course operator error, not poor stove design. However, such fires would be less likely to occur if door latch handles gave unique feedback to the operator only when the door was securely latched. The lack of this special "feel" would warn the operator who was going through the motions of latching the door without actually succeeding.

Airtightness

I use the term "airtight" to describe a wood burner that is so tight that closing the air inlet during a large fire suffocates the fire.

The principal advantage of airtightness in stoves is the convenience of being able to keep a fire burning for a long time. Experiments in my laboratory do not entirely support the notion that airtight stoves are always more energy-efficient than non-airtight stoves; other features tend to be more important. But another attribute of airtightness is the possibility of suffocating a chimney fire by closing the stove, assuming that is the only major source of air for the chimney fire. However, use of airtight stoves tends to result in more creosote accumulation and hence more chimney

Figure 4–2. Thermostatic controls for wood and coal stoves. A bimetallic coil attached to the air inlet damper winds up or unwinds in response to its temperature and thus automatically regulates the amount of air admitted to the fire. The knob on the outside rotates the coil as a whole. Thus setting the knob is like selecting a temperature on an ordinary thermostat.

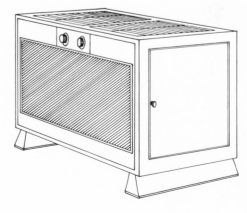

An exposed and simplified thermostatic control mechanism

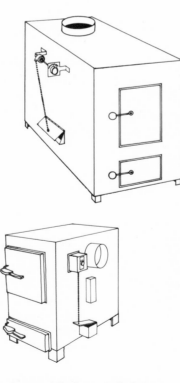

Three thermostatically controlled stoves

93

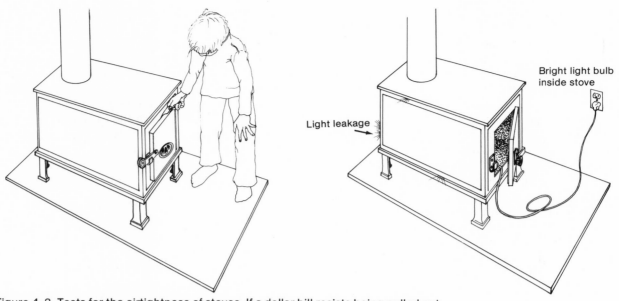

Figure 4–3. Tests for the airtightness of stoves. If a dollar bill resists being pulled out at all locations around the door, the door is probably reasonably tight. For the rest of the stove, air leaks can sometimes be detected via light leaks by darkening the room and putting a very bright electric light inside the stove. It can help to have a second person moving the light around, placing it near the joints being inspected from the outside.

fires. Thus the safety of airtightness can be argued both ways. I personally feel airtightness is desirable for the control it gives. It is the operator of the equipment who must always be responsible for and is usually the cause of creosote buildup. Airtight stoves need not be operated with very restricted air, and thus need not generate undue creosote.

Assessing the airtightness of a stove in a store is difficult, for leaks often are not easy to see. Putting a very bright light inside and looking for light leaking to the outside is one way. If, when a stove's door is closed over a dollar bill, the bill is easy to pull out, the door probably leaks (Figure 4–3).

Asbestos Door Gaskets

The hazard of asbestos in homes has not been adequately quantified, and in particular whether asbestos door gaskets are hazardous. However, it can be difficult for manufacturers to meet worker safety requirements when asbestos is used. Thus substitutes for asbestos may be inevitable in any case. Some doors seal tightly enough with no gasket. The Lange stoves from Denmark are examples. Metal braid is used in some products. Ceramic wool materials such as Kaowool[R] and Fiberfrax[R] (see Appendix 5) may be suitable. (It would also be useful to assess the asbestos hazard of furnace cements used in stove construction.)

Spark Tightness

Sparks should not come out of a wood heater during normal operation. Although proper floor protection should render most sparks harmless, the probability of house fires is higher for wood heaters that emit more sparks.

To minimize the spark emission odds, all *fireplace* stoves should be equipped with effective fireplace screens.

Some ordinary stoves are capable of emitting sparks through their air inlets. Baffled air inlets, with no direct line-of-sight path from the fire to outside the stove, are much less likely to let sparks out (Figure 4–4).

Hot Coals Spillage

Some stoves have a substantial hot-coal spillage problem because the manufacturers specify pulling the coals forward before each refueling, and because there is no significant coal-catching hearth extension. This is perhaps more of an annoyance than a serious hazard since spillage almost always occurs when the operator is present. Accessory coal catchers are available for some stoves with this problem (see Appendix 5).

94

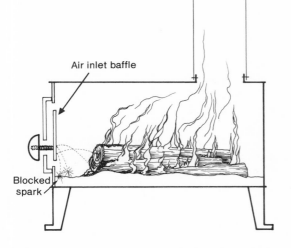

Figure 4–4. A stove with a baffled air inlet on the inside of the door. Such baffles can prevent sparks from flying out through the air inlet. Adequate floor protection (Chapter 2) in front of stoves is an alternative way to minimize this potential hazard. Door baffles also help keep handle temperatures down. Door baffles also affect performance through changing the air flow pattern inside the stove. These effects are not always beneficial. Thus on balance it is not obvious whether door baffles are beneficial.

LISTED WOOD HEATERS

"Listed" wood-heating equipment is increasingly both available and required by state laws. A listed appliance is one that has been examined and tested for *safety*, not performance, by a recognized laboratory, for example, Underwriters Laboratories, Inc. The most reliable way to determine whether equipment is listed is to look for a label on the appliance so stating. Several organizations and laboratories test and list wood-heating equipment, and the standards they use are not always the same. However, the differences are usually small.

What is the significance of listing for a wood heater? First, relatively good and complete installation and operation instructions must be supplied with such appliances. These instructions cover basics such as proper ash disposal and the facts of life about creosote.

Second, these instructions contain safe clearances to combustible walls and floors that have been established experimentally. Thus if the manufacturer's installation instructions are followed, you have a good assurance that normal use, and perhaps even abusive use, of the heater will not overheat the nearby parts of the house. The allowed clearances may be considerably less than NFPA's, which, being general, must be worst-case clearances, safe for the biggest and hottest stoves.

Third, listed wood heaters must conform with certain basic minimum engineering and construction practices. Listed equipment is unlikely to fall apart when used and is likely to last a reasonable time.

The leader in writing safety standards for wood-heating equipment is Underwriters Laboratories, Inc. Established, nationally recognized standards exist for chimneys, factory-built fireplaces, and fireplace stoves. A safety standard for stoves is being developed, and stoves are already being tested and listed according to the preliminary draft of the standard. Standards are also worked on for wood-fired furnaces and boilers and some wood-heating accessories.

Many basic features of the standards for fireplaces, fireplace stoves, and stoves are identical. The following description applies specifically to stoves.

Three tests the Underwriters Laboratories proposed standard for wood stoves involve testing the products under very severe, almost unrealistic conditions. This is appropriate when testing for safety. In *performance* testing for such things as power output and energy efficiencies, being realistic in operating the wood heater is important. But safety testing ought to be abusive; the worst operating habits and hottest fires reasonably imaginable should be used.

There are three parts to UL's fire tests: a radiant fire test, a brand fire test, and a flash fire test. Details vary somewhat, but the following procedures are typical.

Radiant Fire Test

The fuel in the radiant fire test is charcoal briquettes. They are placed in a 6-inch deep grate basket whose bottom is 4 inches above the stove's floor. Every 7½ minutes more fuel is added to keep the grate full, and the ashes on the floor of the stove or in the ash pan are removed. All controls such as air inlets, dampers, and sliding baffles are set so that the stove is as hot as possible. For instance, a thermostatic control on a stove might be bypassed and the air inlet damper held wide open. This intensely hot, radiant fire is maintained for many hours until temperatures of the stove and of the surrounding walls and floor reach steady maximum values. For the stove to pass this test, the wooden walls, floor, and ceiling of the test room may be no more than 117 Fahrenheit degrees hotter than room temperature for exposed surfaces, and 90 degrees hotter than room temperature for covered surfaces, such as the floor under a floor protector. For example, 187° F. is the maximum allowed temperature for exposed wood if the room temperature is 70° F. The test room has only two adjacent walls, a floor, and a ceiling. Thus the heat output of the stove can escape and the testroom air temperature does not become absurdly high.

Since no ash is allowed to accumulate in the bottom of the stove this test is especially severe with respect to floor temperatures. The floor protection used in the test is supplied by the manufacturer.

Brand Fire Test

The brand fire test uses brands for fuel consisting of two crossed layers of zero moisture content, ¾" x ¾" Douglas fir sticks on 1-inch centers (Figure 4–5). Each brand has an area of one-third the floor area of the stove. During the test one brand is added every 7½ minutes. Controls on the stove are adjusted to yield maximum temperatures, and the test is continued for as many hours as it takes for the temperatures of the stove and of the test room structure to stop rising. For the stove to pass this test, no exposed part of the test room may exceed 117 Fahrenheit degrees above room air temperature, and unexposed surfaces must be not more than 90 degrees hotter than room temperature.

Flash Fire Test

The flash fire test is conducted as a continuation of the brand fire test. Eight brands are simultaneously put into the stove. Since the fuel is a resinous wood at zero percent moisture content in very small (¾" x ¾") pieces, well aerated and large in quantity, an extremely hot fire can result (Figure 4–6). The instantaneous heat output rate of even the smallest stoves generally exceeds many hundreds of thousands of Btu per hour during this test. Again, controls are set to maximize temperatures. For the stove to pass this test a somewhat higher temperature rise of the test room structure is allowed: 140 Fahrenheit degrees above ambient, or 210° F. if room air temperature is 70° F.

If a stove fails any of these three tests due to wall temperatures being too high, the manufacturer may either modify the stove design to make the sides cooler, or may specify a larger clearance to walls in the installation manual, after the stove has passed the test at the increased distance from the wall.

In addition to the limits on floor, wall, and ceiling temperatures, other conditions must be met. UL has set maximum temperature limits for the metal parts of the stoves and any electrical components. The metal temperature limitations are not to prevent melting—this is rarely a problem—but are to ensure a certain unspecified durability; high temperatures can cause oxidation and scaling, which eventually will eat through the metal. Temperature limits for the flue gases as they enter the chimney are designed to ensure that

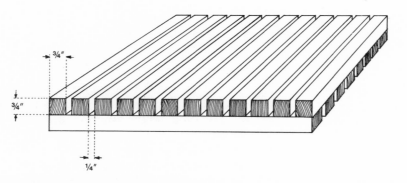

Figure 4–5. A firebrand—the standard fuel used in the safety testing of wood stoves and fireplaces. The brands are constructed of two layers of zero moisture content ¾- X ¾-inch Douglas fir sticks. Each brand is constructed to have ⅓ of the hearth's area. In the "flash fire" test 8 brands are added at once on a bed of hot coals. An extremely hot fire results due to the large amount of fuel, its low moisture content, the small stick size and the air spaces around each stick.

Figure 4–6. The flash fire test in a stove. (Photo courtesy of Fisher Stoves International)

use of UL listed stoves cannot result in higher flue gas temperatures than are used to test chimneys. UL also limits handle temperatures to levels that will not result in burns when touched. Glass doors and panels are given a breakage test. The unit must pass a mechanical stability test by not tipping over under a prescribed push. The unit must also not be damaged when logs are thrown into it (in a "scientific" way).

What "Listed Equipment" Does Not Mean

Users of listed wood heaters should not assume that they are safer than other heaters and require less worry and attention in their installation or operation. Wood-heating equipment is rarely the cause of house fires. It is almost always poor installation or careless operation. No less vig-

ilance is required in installing and operating listed equipment than unlisted equipment.

Underwriters Laboratories very carefully and honestly describes equipment that passes its tests as "listed," not "safe" or "approved." The current safety standards are a conscientious attempt to cover adequately the most critical areas of safety. But many aspects of the tests are necessarily arbitrary on such points as how hot a fire is used, what kind of chimney is used, and how high a temperature of various materials is safe. Certain areas related to safety, such as creosote potential, are not covered at all, except in the appliance instructions.

It is even possible that some stoves may be less safe because of modifications made by the manufacturers to pass the safety test. For instance, limiting the maximum possible opening of the air inlet has the beneficial effect of limiting the maximum temperature of the stove, of nearby combustible materials, and of the flue gases. However, a stove so modified may generate more creosote because of the restricted air supply. Users may also be tempted to leave the door slightly open to get more air to the fire. This may be "necessary" when using the stove at high altitudes or when burning green wood or large chunks of wood. But it is dangerous to leave the door ajar. Thus features providing safety in some areas can create hazards in others. Safety is a matter of degree. No test can prove a product absolutely safe.

Use of equipment that has been tested and listed for safety is safer only if one is just as careful and worried and conscientious about installing and operating the equipment as if it were an unlisted leaky thin-walled Rube Goldberg contraption.

CHAPTER 5

Special Topics

Some definitions of terms used in this chapter.

Furnace. A central heater with hot-air heat distribution.

Boiler. A central heater with either hot-water or steam-heat distribution.

Central heater. An appliance intended to heat a whole typical house uniformly. Virtually all central heaters require pipes or ducts for heat distribution.

Water Heater. A device for heating domestic (or tap) water.

Central Heating with Wood and Coal

Interest is growing in wood-fired furnaces and boilers for heating homes. Because the heat is distributed through conventional means such as forced hot air ducts and baseboard radiators, house temperatures are as uniform as with conventional central heating systems. The wood system is usually backed up by oil, natural gas, bottled gas, or electricity so that heat is automatically assured whenever the wood fire goes out. If a house has a conventional central heater in good condition, it is natural to install an "add-on" or "wood-only" furnace or boiler. "Dual-fuel" systems are also available which burn wood (and/or coal) and a fossil fuel in the same overall appliance.

Many safety considerations for central wood heaters are basically the same as for room wood heaters. Examples are the need for a safe chimney and a sound stovepipe connector, the causes and minimization of creosote, the need for frequent chimney inspection and cleaning whenever necessary, the dangers of back flashes, and the importance of clearances.

Some of these basic areas are even more important for central heaters than for stoves. The larger fuel capacities of central heaters can result in very high creosote accumulation rates. Full loads of fuel on days of only mildly cool weather will result in very smoky, creosote-producing burns. Some manufacturers and importers of wood furnaces and boilers even recommend not burning wood in the early fall or late spring.

This advice indicates how serious the creosote problem can be in central wood heaters. There are ways to deal with it. Creosote accumulation can be minimized despite low heat demand by stoking with only small fuel loads. One or two small hot fires a day can heat a house in mild weather, thus avoiding the slow smoldering, creosote-producing burn. This works particularly well with wood-fired hot-water boilers whose heat storage capacity results in steady house temperatures despite a non-steady fire.

The chimneys used to vent central wood heaters are often subjected to more severe conditions than those serving stoves. The larger heating and fuel capacities of central wood heaters can result in higher stack temperatures, and for longer times. Flue gases can approach 1000° F. for hours at a time. Thus it is vitally important that central wood

Table 5-1. Clearances

NFPA RECOMMENDATIONS — AUTHOR'S ADDITION

	Hand-fired, solid-fuel, warm-air furnaces[1]	Automatically stoker-fired central warm-air furnaces with 250° F. limit control and barometric draft control that cannot be set higher than 0.13 inches of water	Boilers and hot water heaters, all water-walled or jacketed[1]	Conventional oil, gas, or electric furnace with a solid-fuel supplemental furnace discharging hot air into the hot air plenum of the conventional unit
Front	48 Inches[2]	48[2]	48[2]	—[7]
Sides and Rear	18	6	6	—[7]
Above and sides of bonnet or plenum	18	6	6[3]	18
First 3 feet of hot air supply ducts	18	6	—[4]	6
Second 3 feet of hot air supply ducts (duct between 3 to 6 feet from plenum)	6	6	—[4]	1
Supply duct more than 6 feet from plenum	1[5]	1[5,6]	—[4]	1[5]

1. If the appliance is not fully water-walled or fully jacketed, clearances from single-walled sides should be those for radiant stoves.

2. The clearance from the front is not so much larger than the side and rear clearances for safety, but for reasonably convenient access for operation and maintenance.

3. This is the minimum clearance from the top (not the plenum) of boilers and water heaters.

4. The minimum clearance for hot water and steam pipes is 1 inch, unless the system is equipped with a limit control which cannot be set to permit a water temperature above 150° F. Where pipes pass through a floor, wall or ceiling, the clearance at the opening through the finish flooring, wall paneling or ceiling may be reduced to not less than ½-inch. The gap should be covered with a noncombustible material, such as a metal ring.

5. Where a horizontal or vertical duct, riser, boot, box, or register passes through or is installed in a floor, ceiling, wall, or partition, the clearance or air space must be 3/16-inch or the component must be of double-wall construction with an air space of at least 3/16-inch between the two walls. The front or face of registers may be in contact with combustible material. Vertical ducts or risers, and resisters should not be less than 6 feet from the plenum, measured along the horizontal ducts.

6. This minimum clearance drops to zero beyond the point where there has been a change in direction equivalent to 90 degrees or more.

7. The minimum clearances from the walls of the conventional furnace itself are not affected by this kind of add-on furnace installation. The add-on furnace and the hot air duct from it to the plenum of the conventional unit should have 18 inches of clearance to combustibles.

Minimum clearances to combustible materials from solid fuel and dual (or combination) fuel central heating systems. This information is adapted in part from NFPA 89 M, "Heat-Producing Appliance Clearances 1976," and 90 B, "Warm Air Heating and Air Conditioning Systems 1976." Reduced clearances are allowed with appropriate protection, as in Table 5-2. The duct clearances are for uninsulated single-wall ducts.

heaters be vented into safe chimneys. The basic standards for safe chimneys are identical for all solid-fuel appliances. But with large central wood heaters it is especially important to avoid use of masonry chimneys without liners and without the proper clearances between the chimney and combustible parts of the house. Although connecting separate wood and oil or gas appliances to the same flue is not recommended, if done it is vital to keep the flue clean.

Clearances

Required clearances from combustible materials are often specified in the installation instructions furnished with furnaces and boilers. If not, the unit should be installed with the clearances in Table 5–1. For normal boilers, fully enclosed in water walls and top except at access doors and flue collar, the minimum clearance to combustibles is

Table 5-2. Reduced Clearances

Clearances, in inches, for Furnaces, Boilers, Plenums, and Ducts with Specified Forms of Protection, According to NFPA[1,2,3,4]

TYPE OF PROTECTION

Applied to combustible material unless otherwise specified and covering all surfaces within the distance specified as the required clearance with no protection. Thicknesses are minimal.	WHERE THE REQUIRED CLEARANCE WITH NO PROTECTION IS:			
	18 Inches		6 Inches	
	Above	*Sides & Rear*	*Above*	*Sides & Rear*
¼-inch asbestos millboard spaced out 1 in.[2]	15	9	3	2
0.013-inch (28-gauge) sheet metal on ¼-in. asbestos millboard	12	9	3	2
0.013-inch (28-gauge) sheet metal spaced out 1 in.[2]	9	6	2	2
0.013-inch (28-gauge) sheet metal on ⅛-inch asbestos millboard spaced out 1 in.[2]	9	6	2	2
¼-inch asbestos millboard on 1-inch mineral wool batts reinforced with wire mesh or equivalent[4]	6	6	2	2
0.027-inch (22-gauge) sheet metal on 1-inch mineral wool batts reinforced with wire or equivalent[4]	4	3	2	2
¼-inch asbestos millboard	18	18	4	4
¼-inch cellular asbestos	18	18	3	3

1. All clearances should be measured from the outer surface of the appliance or duct to the combustible material disregarding any intervening protection applied to the combustible material.

2. Spacers should be of noncombustible material.

3. Asbestos millboard referred to above is a different material from asbestos cement board. It is not intended that asbestos cement board be used in complying with these requirements when asbestos millboard is specified.

4. The mineral wool batts should have a minimum density of 8 pounds per cubic foot for a minimum melting point of 1500° F.

6 inches. For most hot air central heaters the minimum clearance is 18 inches. Reduced clearances are allowed with appropriate protection of combustible walls or ceilings, as specified in Table 5–2.

Floor protection is not required if the unit is on a concrete slab-on-grade or other fully noncombustible floor, although elevating the unit a few inches above the floor on concrete blocks is generally advisable to keep the unit dry in the event of minor flooding. If the floor is combustible, NFPA's recommended protection is indicated in Table 5–3. As always, for *listed* equipment, manufacturer's instructions supersede NFPA guidelines.

It can be difficult to determine where in Table 5–3 a particular central heater fits. A floor protector that is safe for almost all equipment in use today and is both inexpensive and readily available is sheet metal plus hollow masonry (entry in

Table 5.3. Floor Protection

Type of Protection	Required for the Following Types of Heaters and Furnaces
No Floor Protection:	Residential-type furnaces so arranged that the fan chamber occupies the entire area beneath the firing chamber and forms a well-ventilated air space of not less than 18 inches in height between the firing chamber and the floor, with at least one metal baffle between the firing chamber and the floor.
Asbestos and Metal: A sheet of ¼-inch asbestos covered with a sheet of metal not less than No. 24 U.S. gauge.	Heating furnaces and boilers in which flame and hot gases do not come in contact with the base and which are set on legs which provide not less than 4 inches open space under the base.
Hollow Masonry: Hollow masonry not less than 4 inches in thickness laid with ends unsealed and joints matched in such a way as to provide free circulation of air through the masonry.	Downflow furnaces.
Hollow Masonry and Metal: Hollow masonry not less than 4 inches in thickness covered with a sheet of metal not less than No. 24 gauge. The masonry must be laid with ends unsealed and joints matched in such a way as to provide a free circulation of air from side to side through the masonry.	Heating furnaces and boilers in which flame and hot gases do not come in contact with the base.
Two Courses Masonry and Plate: Two courses of 4-inch hollow clay tile covered with steel plate not less than 3/16-inch in thickness. The courses of tile must be laid at right angles with ends unsealed and joints matched in such a way as to provide a free circulation of air through the masonry courses.	Heating furnaces and boilers in which flame and hot gases come in contact with the base.

Presumably, these protectors are required only directly under the appliance, but in addition, according to NFPA, a sheet of ¼-inch asbestos covered by a sheet of metal not less than No. 24 U.S. gauge is required extending at least 18 inches from the appliance on the front side where ashes are removed. (The sheet of asbestos may be omitted where the protection required under the appliance is a sheet of metal only.) If the appliance is installed with clearance less than 6 inches, the protection for the floor should be carried to the wall.

Floor protection for central heaters according to NFPA. (This table is adapted from Table 7-3H on p. 7-55 of *Fire Protection Handbook,* 14th Edition, National Fire Protection Association, Boston, Mass., 1976.)

Table 5–3 and Figure 2–58). The masonry itself normally need only extend under the appliance itself, but some form of protection against sparks and embers is required at least 18 inches beyond the sides with doors. This protection should extend 18 inches to the sides as well as in front of the door. If the appliance is also a radiant heater, floor protection should follow the guidelines in Chapter 2 for radiant heaters.

Overheat Control

Central solid-fuel heating systems are most hazardous when overheated. This is of course true of room heaters also, but codes and standards generally require much more in the design and installation of central heaters to prevent or deal with overheating. In the case of boilers this is partly because explosions are an added and very serious potential problem. In addition, many furnaces and boilers are double-walled structures, are insulated, and normally depend on pumps or blowers to take heat away from the unit (and distribute it to the house). For these reasons temperatures at some locations can be very high during certain kinds of malfunctions. Thus many central heaters are more susceptible to overheat damage than are most stoves.

Boilers can have as many as four overheat preventative features. Listed in the approximate

102

order in which they would come into play, they are:

1. A control that closes the air inlet damper as the boiler heats up.

2. An overheat control that dumps heat into the house when the boiler is too hot regardless of whether the house needs heat.

3. A pressure relief valve that vents steam or hot water out of the boiler.

4. A fusible plug in the firebox wall that melts, letting boiler water out to douse the wood or coal fire.

Furnaces can have three overheat preventative features:

1. A fan control that forces heat into the house whenever the air in the plenum is hot, again regardless of whether the house needs heat.

2. A combustion air control that stops air from feeding the fire whenever the plenum is overheated.

3. A fusible-link or bimetallic heat-dumping panel on the plenum that opens when temperatures are high. Care must be taken that such a heat-dumping panel does not create a worse hazard, that the dumped hot air and radiation do not themselves start a fire.

Even in systems equipped with all the appropriate safety features, overheating is to be avoided. It is at least inconvenient to have some of the more extreme features used, such as the melting of fusible plugs or links. Avoidance of overheating problems involves both selecting the right equipment and careful operation.

The most critical aspect of equipment selection is sizing. Central heaters with more heating capacity than a house needs are much more likely to overheat. Boilers with relatively small water volumes are also susceptible. Careful operation involves not overloading the unit, not using particularly flammable fuel, not leaving the unit's doors open, not abusing the thermostatic combustion air control—the same things that are important for not overheating stoves.

Power Failures

Solid-fuel central heating systems must be designed to be safe when electric power fails. Since most systems use electric pumps or blowers to move the heat from the furnace or boiler to the living spaces of a house, heat builds up substantially in the central heater when electric power is lost. This heat can damage the unit and can cause dangerous pressure buildups in boilers and dangerously high plenum and duct temperatures in hot-air systems.

Minimizing the extent of the heat buildup during a power failure requires stopping the combustion-air flow to the fire. This can be done manually if an operator is present. But automatic shutdown should be incorporated in the design. Many central heaters have thermostatically controlled combustion air systems that automatically stop the air flow when the heated water or air reaches a certain temperature. This provides some protection against overheating during power failure as well as during normal operation but it is better still to stop the combustion air at the moment of the power failure, *before* the temperature of the system begins to rise. Most electrically powered combustion air controls do this— they shut whenever they lose electric power. Purely mechanical controls do not. Bimetallic thermostats and aquastats will not shut combustion air dampers until the temperature rises. Spring- or gravity-loaded electric override could be incorporated in these systems that would force the damper shut in case of a power failure.

Wood central heating systems can be designed to work without electric power. Conventional ducts or piping are usually not very effective and are impractical to modify. Extra large diameter ducts or pipes are needed, and they should everywhere be sloping upward from the central heater. Water systems should have an automatic air release valve at every high point in the system to let out air. Such thermosyphon or "gravity flow" systems were not uncommon 50 years ago. The older editions of heating engineering handbooks had considerable information on gravity flow heat distribution systems.

Great care is necessary when using a contemporary solid-fuel central heater during a power failure. Fires should be small and temperatures and pressures inside and in the vicinity of the burner and its heat distribution network should be closely monitored. In hot-water systems, flow check valves and thermostatically controlled electric valves must be manually opened to allow the hot water to flow.

Dual-Fuel Central Heaters

Both furnaces and boilers are available that include a gas or oil or even electric capability as well as the solid fuel. Whether a dual-fuel unit, or two separate units, one for each fuel, is better is principally an economic question. Both systems can be made to function effectively.

A safety issue concerns venting the flue gases. Most dual-fuel central heaters have a single flue collar and thus both the solid fuel and the oil or gas combustion products will necessarily be vented into a common chimney flue. Thus most of the safety concerns discussed in Chapter 3 concerning multiple use of single flues apply—the need for adequate capacity, possible blockage of the flue with creosote in the chimney, in the connector, and with fallen creosote at the breaching, the added difficulty of suffocating a chimney fire, and the dangers of small explosions. Keeping the chimney, stovepipe, and cleanout area clean will eliminate many of these hazards. With a good initial installation and conscientious maintenance, such installations are reasonably safe.

Separate Combustion Chambers

Dual-fuel central heaters are generally considered safer if the solid fuel and fossil fuel are burned in separate combustion chambers. However, some of the objections to single-combustion-chamber units can be met through careful design. The fossil fuel burner must be located or protected so that it cannot be hit with logs. This can be done by recessing the burner into an alcove out of the solid fuel chamber and/or by placing it high in the firebox. The fossil-fuel burner, including its ignition system, should be protected against fouling with creosote. A small amount of intentional air leakage around the burner will keep the wood smoke away. Single-combustion-chamber units should be designed to halt the fossil fuel burning whenever the refueling door is open.

Both separate-chamber and single-chamber dual-fuel systems can have small explosions if ignition of the fossil fuel is delayed, due, for instance, to creosote-fouled electric ignition systems. Flaming or glowing wood can be the ignition source. Both the fouling and this ignition may be more likely in a single-chamber unit. Such explosions are created intentionally during testing of such units to assure the units will not be damaged. However weak chimneys may be damaged, and sparks may be blown out of other appliances that are connected to the same flue.

Use of separate combustion chambers can eliminate or reduce some of these hazards, but so

also can good engineering in a single chamber unit. I do not see compelling evidence in favor of either design type.

Furnaces (Hot-Air Central Heaters)

An area of special concern with solid-fuel *furnaces* is preventing wood joists and flooring close to the hot air ducts and the plenum from overheating. Many wood- and coal-fired furnaces are capable of generating hot-air temperatures much higher than the normal 250° F. maximum. A number of house fires have been caused by this hot air circulating through existing ducts not designed for such high temperatures. Clearances must be designed for the worst case, which is usually a roaring hot fire during a power failure.

The minimum clearances recommended by NFPA are indicated in Table 5–1 and Figure 5–1. These clearances apply to furnaces using wood or coal and to dual-fuel or combination furnaces when one of the fuels is solid. Since supplemental furnaces are a relatively new development, NFPA in 1979 had not completed recommended clearances for this case. Clearances around the supplemental unit itself are covered in Column 1 of Table 5–1, but since these units lack their own plenums and ductwork, it is not obvious what clearance should apply to these parts of the existing conventional furnace. Most wood furnaces can produce much hotter air than an oil- or gas-fired unit. Thus the clearances from the existing ducts and plenum are usually inadequate with the add-on furnace installed.

In the common and recommended installation method of connecting wood furnace hot air output into the hot-air plenum of the conventional furnace, I recommend observing the plenum and duct clearances in Column 4 of Table 5–1. In most installations these clearances do not already exist and are difficult if not impossible to achieve. Reduced clearances are safe with adequate protection as outlined in Table 5–2. But since space around ducts is often limited, even adding protection is often very difficult. In many cases the duct system must be rebuilt. Special double-walled or insulated ducts may be necessary. Use of UL-listed Type L, Type B, or Type B-W vents makes safe 1 inch of clearance for the first 6 feet of duct. These "vents" are designed as chimneys for certain types of gas and oil appliances.

If homemade insulated ducts are attempted, the following three points should be observed:

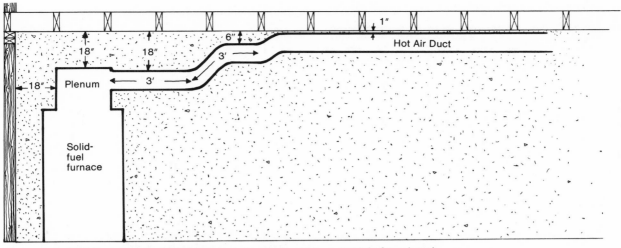

Figure 5–1. NFPA-recommended clearances to combustible materials from hand-fired wood or coal furnace plenums and horizontal ducts. Vertical ducts or risers may be installed starting 6 feet from the plenum and should have 3/16-inch clearance from all combustible materials, or be of double-wall construction with a 3/16-inch air space between walls.

1. The insulating material must be noncombustible. Ordinary fiberglass will work, but insulation designed for higher temperatures is much better, especially for use close to the solid-fuel furnace.

2. If paper- or foil-backed insulation is used, the backing must not contact the duct. The foil backing on much fiberglass insulation *is* combustible since it is foil applied to paper.

3. There should everywhere be at least a 1-inch free air space between the insulation and combustible materials. The insulation should *not* come in contact with the wood joists.

Manufacturer's instructions should explain appropriate kinds of hookups and the necessary controls. Two important safety principles involved are avoiding negative pressure inside any furnace jacket so that combustion products cannot be sucked out of leaks into the house air stream, and assuring safe temperatures during power failures. In the event of a power failure, the combustion air damper should close and the layout of the system should ensure that residual heat does not enter the return ducts but flows towards the plenum and supply ducts, where clearances and protection are more likely to be able to handle the high temperatures.

The best installation type is parallel, with direct (ducted) feed of return air to the wood furnace (Figure 5–2). Series installations are only suitable if the wood furnace comes second, and only with large wood furnaces and interconnecting duct work so that air flow through the whole system is not unduly restricted (Figure 5–3).

Series connections with wood furnace coming first (Figure 5–4) where it acts as a preheater for the conventional furnace is the least satisfactory arrangement for the following reasons:

1. With high temperature air bathing the blower motor in the conventional furnace, the motor's life expectancy is reduced.

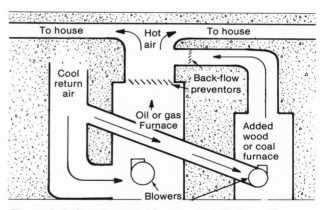

Figure 5–2. Supplemental furnace installation with parallel connection and direct (ducted) feed of return air to the wood furnace. This is the preferred installation for supplemental solid-fuel furnaces. The two back-flow-preventing dampers assure that when either unit is operating by itself the hot air will indeed get into the house and not be short-circuited through the other unit. Some attention must be given to pressure matching to assure that when both blowers are on, the pressures developed in one unit do not inhibit air flow through the other.

105

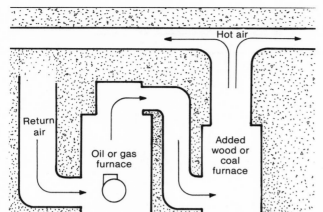

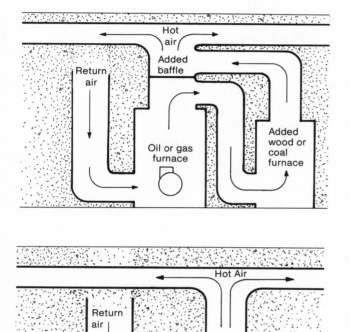

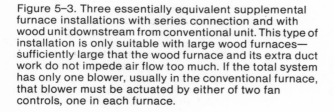

Figure 5-3. Three essentially equivalent supplemental furnace installations with series connection and with wood unit downstream from conventional unit. This type of installation is only suitable with large wood furnaces—sufficiently large that the wood furnace and its extra duct work do not impede air flow too much. If the total system has only one blower, usually in the conventional furnace, that blower must be actuated by either of two fan controls, one in each furnace.

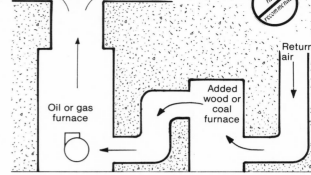

Figure 5-4. Supplemental wood furnace installation with series connection and with wood unit upstream. This type of installation is not generally advisable despite its simplicity.

2. The high temperatures may also trip the thermal overload switch in most blower motors, turning off the blower. (These first two problems do not arise if the blower motor is outside the air flow.)

3. The blower in the conventional furnace may create negative pressures or suction inside the wood furnace's jacket. This will draw wood smoke out of any cracks or leaks, and this smoke will then be blown into the house.

4. Reverse air flow is more likely during power failures or blower failures, resulting in dangerously high temperatures in the return air ducts.

Systems in which the cold air input to the supplemental furnace comes directly from a basement or utility room space, instead of a ducted return air system, can lead to difficulties and is prohibited by some local building codes. The air circulated to the house may be laden with dust and basement odors. Also the energy efficiency of such installations tends to be lower.

Perhaps most important from a safety point of view is that if the supplemental furnace has a relatively large blower or the installation room is relatively tight, the utility room may be slightly depressurized when the blower is on. This will reduce the draft for all fuel-burning appliances in the room. The effect will be most on natural draft appliances such as the add-on furnace and gas appliances. In extreme cases the consequences could be very serious. Possibly toxic flue gases from a gas water heater or furnace may not get up the chimney but be pulled into the utility

106

room and forced into the house. Providing ventilation to the utility room either from the house or the outdoors will usually solve the problem where it does arise. But in general it is better to duct return air directly to the supplemental furnace.

Boilers

Most wood and coal boilers do not *boil* water but only heat it; despite the contradiction in terms such units are called "hot water boilers." Residential solid-fuel steam boilers are available, but the demand is not high since residential steam heat distribution systems have not been popular for many years. As is the case with furnaces, both solid-fuel-only heaters and dual-fuel units such as those burning wood and oil are available. The solid-fuel-only units are usually used to supplement an existing conventional boiler.

Any system that heats water is potentially very dangerous. Since most of the dangers and pre-

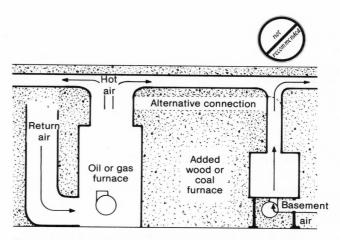

Figure 5–5. Supplemental furnace installation with no *ducted* return air input. The wood furnace draws its heat-transfer air from the utility room or basement and can cause depressurization of tight utility rooms with consequent reduced drafts for all fuel-burning appliances, and even chimney flow reversal and *lack* of venting. Such installations should either be avoided or attempted with caution.

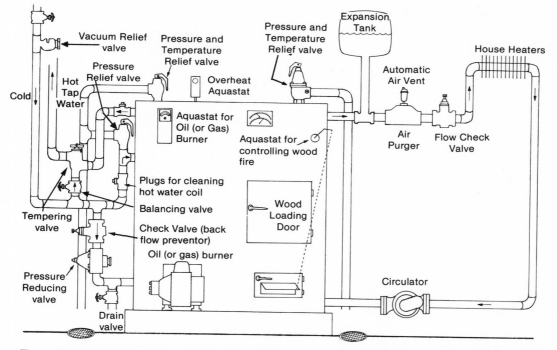

Figure 5–6. A dual-fuel boiler installation including domestic hot water. A dual-fuel system must have enough chimney capacity to handle both fuels being used simultaneously. It must be able to dump heat automatically if the boiler starts to overheat. This may preclude use of self-contained non-electric zone valves. In simple systems the overheat aquastat might turn on the circulator when the boiler water exceeds about 200° F. even if the house or zone does not need heat. The vacuum relief valve must be higher in elevation than the boiler. In the event of loss of supply-water pressure, this valve prevents loss of water via siphoning by letting air into the piping. This valve can also prevent collapse of some types

of tanks. Most building codes require a pressure and temperature relief valve for the domestic hot water tank, or just a pressure relief valve if domestic water is heated in a coil. The air purger and vent prevent air locks from forming and eliminate much of the noise caused by air being in the system. The tempering valve is placed lower than the hot water outlet from the boiler to prevent hot water from rising into the cold water pipes. The balancing valve is to compensate for possible flow resistance in the water heating coil. The flow check valve is to prevent thermosyphon circulation when the circulator is off. The size of the plumbing has been exaggerated for clarity.

ventive measures are the same if the water will be used for space heating or hot tap water, the major safety issues relating to boilers are discussed together with those of water heaters later in this chapter.

Dual-fuel boilers are installed essentially just as are conventional boilers (Figure 5–6). The house thermostat usually controls the circulating pump. One aquastat (a thermostat for water) controls the oil or gas burner, and another aquastat, set at a lower temperature, controls the flow of combustion air to the solid fuel. Thus when solid fuel is not added the fossil-fuel burner automatically turns on.

An add-on boiler supplementing an existing fossil fuel boiler can be installed in two basic ways—series or parallel (Figures 5–7 and 5–8). Both work satisfactorily. The parallel arrangement has some advantages, but usually costs more. When the wood boiler is not in use, hot water from the fossil fuel boiler is not circulating through it, so heat is not wasted out its chimney or through its walls. If appropriate valves are installed, either unit may be operated independent of the other. Parallel systems are usually more

adaptable to gravity flow heat distribution. If the conventional boiler contains a coil for domestic hot water, the wood boiler can be used for domestic hot water without the house being heated at the same time. One or two small hot fires a day can provide hot water, even in the summer.

A variation available in any boiler installation is a thermostatically controlled mixing valve instead of a thermostatically controlled circulator (Figure 5–9). In this case the circulator is on all the time during the heating season and the thermostat controls the *temperature* of the circulating water by mixing hot water from the boiler with varying amounts of return water from the house's radiation. As a result heat tends to be drawn more steadily from the boiler. Since wood fires cannot turn on and off suddenly, this steady demand for heat is a more natural match to solid fuel boilers. The chances of the boiler overheating are slightly reduced and perhaps also the amount of creosote accumulation. There is also less noise in the radiators or convectors since the water temperature is steadier. There may also be less dust blockage in baseboard finned tubing due to replacement of occasional higher velocity air con-

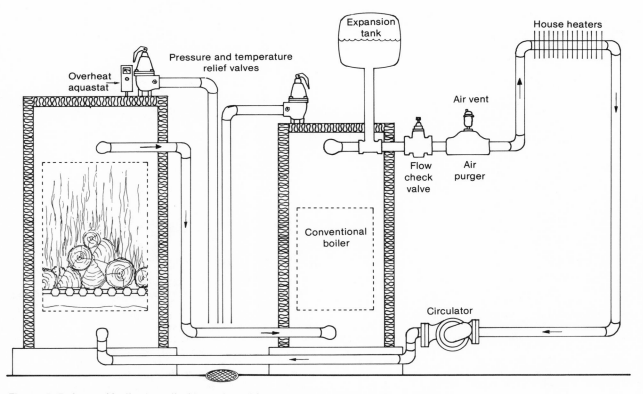

Figure 5–7. A wood boiler installed in series with a conventional boiler. The overheat aquastat turns on the circulator when the wood boiler starts to get overheated, at about 200° F., even if the house needs no heat.

vection with steady but lower velocity convection. There are also claims of lower oil or gas consumption when the conventional burner is used. The extra annual cost of the electric power to have the circulator on all the time is $10 to $30, a cost I consider small compared to the advantages of the system. Thus my recommendation for installing a supplemental boiler is a parallel arrangement (Figure 5-8) with a thermostatically controlled mixing valve.

Boilers need provisions for dumping heat. Frequent opening of a pressure relief valve should be avoided. The boiler room floor will get wet if the valve itself does not discharge over a drain. The valve itself may get solid material stuck in it and not fully close, resulting in constant dripping. Frequent loss of water from a boiler will also result in more interior corrosion due to dissolved oxygen and minerals in the make-up water. For these reasons and added safety, a second aquastat should be installed in the boiler to override the house thermostat and turn the circulator or a zone valve, or the mixing valve on when the boiler starts getting dangerously hot (around 190° to 220° F.) whether or not the house needs heat.

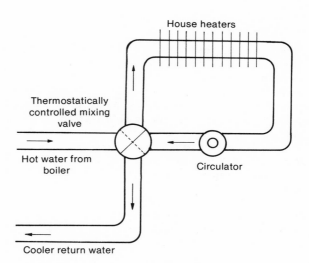

Figure 5–9. A hot-water heat distribution system using a mixing valve. The house thermostat controls the position of the valve, causing variable amounts of return water from the house to be recirculated, thus controlling the temperature of the circulated water. Advantages of mixing valves include steadier demand on the wood boiler and quieter heating in the house, due to less thermal expansion and contraction.

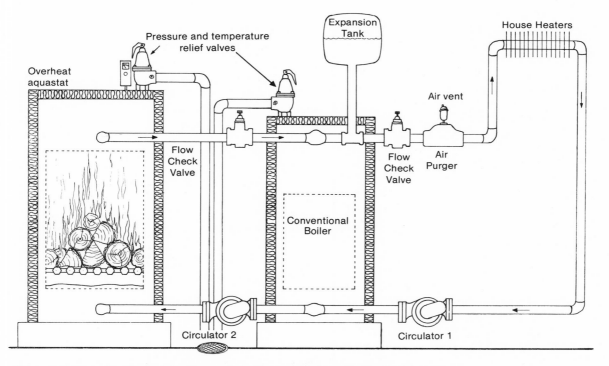

Figure 5–8. A wood boiler installed in parallel with a conventional boiler. The circulator on the right (1) is controlled by the house thermostat. The other circulator (2) is controlled by an aquastat in the wood boiler. It circulates the wood-heated water to the conventional boiler where it can heat domestic water as well as be circulated into the house. The overheat aquastat turns on circulator (1) when the wood boiler water exceeds about 200° F., regardless of whether or not the house needs heat.

Water Heaters

Note: Heating water with wood has more subtle dangers than does heating air with wood. I do not recommend construction of homemade systems without consulting a competent professional plumber with experience in solid fuel water heaters. The information in this section is not intended to encourage you to build your own system, but rather to describe the hazards and to increase the safety of those homemade systems that have already been or will be built despite any advice to the contrary.

Nationwide, about 13 percent of total home energy use is to heat domestic water. About 20 percent of the total energy used for both space and water heating is for the latter. Thus heating water with wood is a natural next step beyond space heating with wood for energy-conscious people.

As more homes become equipped with solar water heating, wood water heating becomes attractive as a natural complement, supplying heat during cold or cloudy and cool periods when the solar system's output is inadequate. A common storage tank can be used for both solar- and wood-heated water.

Wood water-heating systems are generally combined with conventional water heaters to assure a good supply of hot water whether or not there is a wood fire. The wood system acts as a preheater for the conventional system. Use of the wood system saves fuel or electricity costs for the conventional water heater. Note however, that a conventional water heater's guarantee may be voided by adding a wood or solar preheater.

Water heating with solid fuel is most often thought of as an adjunct to space heating. Thus the heat exchanger (the coil, tank, or piping at the heat source, Figure 5-10) is usually an accessory or an add-on to a stove or range or central heater. However solid-fuel burning units are available that are designed *principally* for water heating. These units are of interest in warm climates for year-round water heating, and for swimming pool heating. Swimming pool heating may have special problems. The very low boiler water temperature can result in substantial creosote in the boiler, and the chlorine and other chemicals in the water may cause some corrosion problems.

Some wood-plus-conventional water heating systems are illustrated on the following pages. These systems are reasonably practical and safe unless indicated otherwise on the figure. Entirely different systems, and different combinations and arrangements of components are possible. A competent plumber should be consulted on every installation.

The following are some design considerations related principally to the useful and efficient *functioning* of water-heating systems (safety issues are discussed later):

1. *Heat exchanger location.* Water heat-exchanger surfaces are necessarily relatively cool and thus accumulate more creosote than do other nearby surfaces. The creosote insulates the heat exchanger, reducing the amount of heat transfer, and the creosote deposits are often very difficult to remove. Thus cool, smoky locations, such as inside a stovepipe connector, should be avoided (see Figure 3–8h). When placed inside the combustion chamber next to the fire, heat transfer surfaces tend to be "self-cleaning." Although some creosote accumulates, it is rarely enough to seriously inhibit heat from getting through because the creosote periodically burns off.

Another suitable location for water heat exchangers is outside the combustion and flue-gas system. This avoids creosote fouling, but since the space is cooler, a larger surface area heat exchanger is needed. Simply placing a large uninsulated metal tank near a radiant stove will preheat water significantly. A copper coil around the outside of the stovepipe can function well. For optimum efficiency, the surfaces towards the stovepipe should be blackened, and the surfaces pointing towards the room should be left as bare clean copper. Wrapping insulation around the outside will improve the efficiency but can be hazardous unless stainless steel stovepipe is used—the insulation can result in dangerously high temperatures for ordinary stovepipe, possibly leading to burnout, particularly during a chimney fire.

In furnaces and circulating stoves, suitable locations for water heat exchangers are between the jacket and the combustion chamber wall, or in the hot air plenum.

2. *Hot water storage.* Almost all wood heating systems require a hot water storage tank. If water flows through the heat exchanger only when a hot water faucet is turned on, the water will not get very warm with only one fast passage through the heat exchanger. The water in the heat exchanger will be boiling when no faucet is open, creating a potentially very serious hazard. Also, hot water would be available only when the fire is hot.

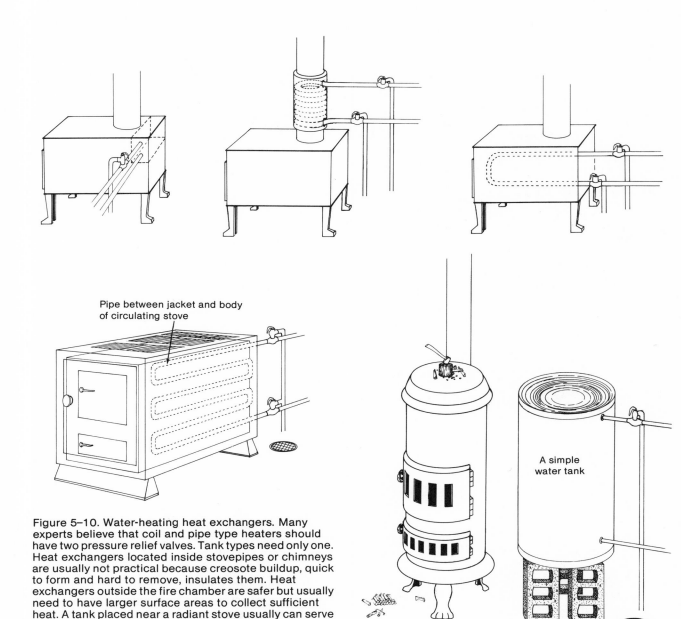

Figure 5–10. Water-heating heat exchangers. Many experts believe that coil and pipe type heaters should have two pressure relief valves. Tank types need only one. Heat exchangers located inside stovepipes or chimneys are usually not practical because creosote buildup, quick to form and hard to remove, insulates them. Heat exchangers outside the fire chamber are safer but usually need to have larger surface areas to collect sufficient heat. A tank placed near a radiant stove usually can serve only as a gentle but significant pre-heater.

Pipe between jacket and body of circulating stove

A simple water tank

A hot water storage tank permits water to flow constantly through the heat exchanger in the fire and pick up heat, and provides hot water for hours after the fire has gone out. The storage tank can be the tank of conventional gas, oil, or electric water heater used as a backup, or a separate storage tank, or the wood water heater itself if it has a very large water capacity. Fifty to 150 gallons is a useful amount of hot water storage.

3. *Circulation.* Water should circulate through the heat exchanger whenever the fire is hot. A temperature-controlled electric pump can be in-stalled to do this job. In some cases natural circulation, powered by the wood heat, can be used.

Pumpless circulation requires that the storage tank be at least level with, but preferably higher than, the heat exchanger (Figure 5–11). Hot water is less dense than cool water and hence tends to rise (Figure 5–12).

Terms used to describe such systems are *thermosyphon* and *gravity flow*. Thermosyphon systems are noiseless and reliable when properly designed. Power failures do not affect them. Water only circulates when it should, when the

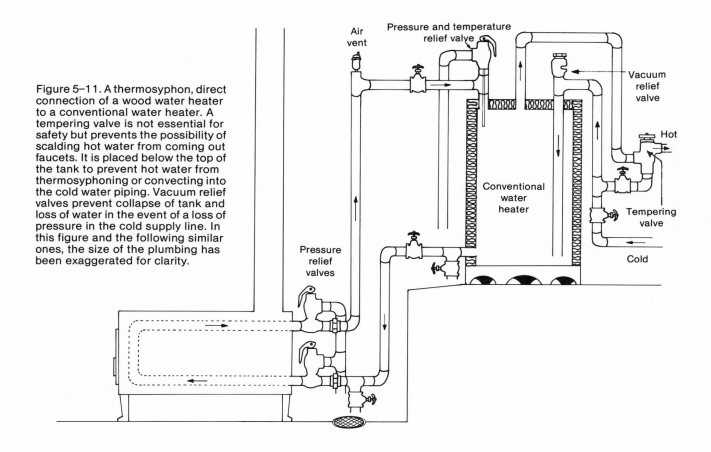

Figure 5–11. A thermosyphon, direct connection of a wood water heater to a conventional water heater. A tempering valve is not essential for safety but prevents the possibility of scalding hot water from coming out of faucets. It is placed below the top of the tank to prevent hot water from thermosyphoning or convecting into the cold water piping. Vacuum relief valves prevent collapse of tank and loss of water in the event of a loss of pressure in the cold supply line. In this figure and the following similar ones, the size of the plumbing has been exaggerated for clarity.

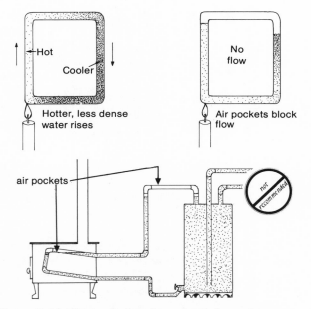

Figure 5–12. Thermosyphon (or natural circulation, or gravity flow) principles, and air-pocket flow-blockage. All piping must slope upwards, or air must be able to get out at high points in the system. Also helpful are large diameter piping, minimum numbers of elbows and other fittings with flow resistance, and at least as much vertical rise as horizontal run in the piping.

water in the heat exchanger is hotter than the water in the storage tank. In fact, the rate of the flow increases with increasing temperature differences, a valuable feature not available in most pumped systems.

Since the forces for the natural circulation are very weak, resistance to flow must be minimized. This usually means using piping roughly 50 percent larger in diameter than with pumped systems, keeping it as short as possible, especially horizontal runs, and using few elbows and only relatively free-flowing isolation and bypass valves such as gate valves. Insulating the hot line will increase flow. Insulating both lines will conserve the collected heat.

There must be no places in the piping where air could get trapped and block thermosyphon flow. Purging the system of air once at the beginning of each season is not adequate since the air dissolved in all water is driven out of the water by the heat. Air locks can be prevented by having all piping slope upwards towards the storage tank or towards automatic air release valves. This permits air bubbles to rise up and out of the thermosyphon

112

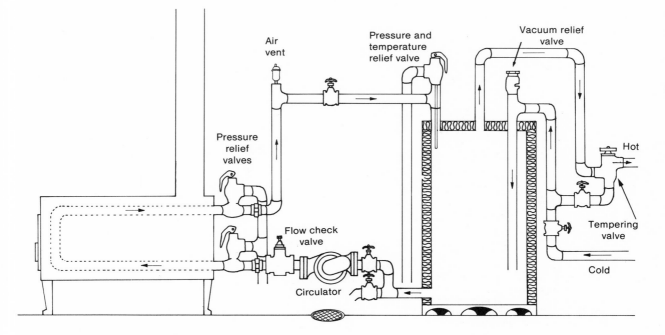

Figure 5–13. A side-by-side direct connection of a wood water heater to a conventional water heater, with electrically powered circulation. The circulator should be on whenever there is useful heat to be collected. The circulator may be controlled by a thermostat in the stovepipe, a surface thermostat on the stovepipe, stove or hot water pipe, or an aquastat in the hot water pipe.

piping. If possible, "horizontal" segments of piping should have a slight slope.

If the storage tank and heat exchanger are side by side, natural circulation will not be as vigorous. In addition, when the wood burner is not in use, circulation may reverse, with heat from the storage tank being lost into the cold stove and perhaps up its chimney. An automatic one-way or flow check valve can prevent this backflow. It must open with a *very small* water force since the driving force of thermosyphon flow is so weak. If the valve ever sticks shut, the water in the heat exchanger may boil. The one-way oil-floating-on-water valves used in some solar heating systems may be suitable, but are not readily available. I do not recommend side-by-side thermosyphon systems.

If the water in the heat exchanger can be heard boiling, the water circulation rate is too low, the heat exchanger is too efficient, or the storage capacity is inadequate. Boiling is a sign of overheating and should be avoided.

For all arrangements of heat exchanger and storage tank, a water pump will guarantee good circulation (Figures 5–13 and 5–14). The pump can be thermostatically controlled. The system should be designed to avoid frequent switching on and off of the pump. It is still desirable to prevent reverse circulation, but the one-way valve need not be so sensitive. In case of power failures, fires should be kept small enough to avoid boiling the water in the heat exchanger.

4. *Separation of hot water storage from backup heater.* There are some advantages in having a second storage tank in addition to the tank of the conventional backup water heater (Figure 5–15). The second storage tank can enable the wood system to supply more heated water. For example, if the thermostat in the conventional water heater is set at 150° F., wood-heated water at any temperature less than 150° F. cannot be used. Thermosyphon circulation will not occur for heat exchanger water at temperatures less than storage tank temperatures, and if a pump forces circulation, the water in the storage tank will be cooled.

This problem can be solved with a second

113

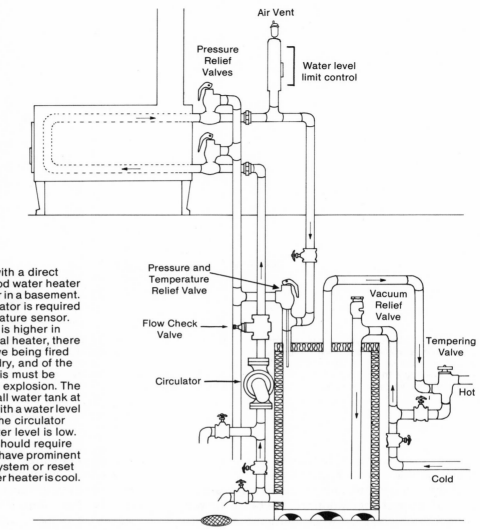

Figure 5–14. An installation with a direct connection of an upstairs wood water heater to a conventional water heater in a basement. An electrically powered circulator is required and is controlled by a temperature sensor. When the wood water heater is higher in elevation than the conventional heater, there is a higher chance of the stove being fired when the heat exchanger is dry, and of the circulator then coming on. This must be prevented to avoid a possible explosion. The method illustrated uses a small water tank at the high point of the system, with a water level limit control which prevents the circulator from turning on when the water level is low. The water level limit control should require manual resetting, and should have prominent instructions not to refill the system or reset the control until the wood water heater is cool.

storage-only or tempering tank between the wood burner and the backup water heater. The capacity of this tank should be about as large as that of the backup heater, typically 50 gallons or more. With the cold water supply entering this tank at 45° to 55° F., even if the wood system heats it up only to 100° F., about half the energy needed has been contributed by the wood system. For the same reasons such second tanks are also useful in solar water heating systems, and in combination with wood and solar systems. The tank of course should be insulated for most efficient operation.

This system has other advantages as well. It increases the system's heat storage capacity. Thus more wood heat can be stored and used, and the household is less likely to run out of hot water. This extra storage tank often makes it easier to do without a circulating pump. If the tempering tank can be installed above and not too far from the heat source, the water can thermosyphon between these two components. In the rest of the system the well pump or town water pressure pushes the water. Thus an electric pump is not needed regardless of the relative locations of the wood water heater and the backup water heater. However, using a pump may be cheaper since cost of the tank and extra piping is significant.

5. *Closed-loop or recirculating heat exchangers.* There can be advantages to systems that do not circulate the actual tap water directly

114

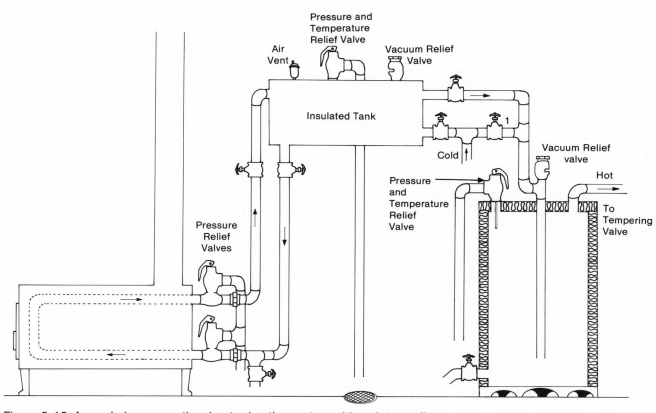

Figure 5–15. A wood-plus-conventional water heating system with an intermediary storage tank. This type of installation makes it possible to use natural circulation even when the conventional water heater is far away or lower in elevation than the wood heater. Also, lower temperature, wood-heated water can be utilized which can increase the contribution of the wood system. When the wood heater is not functioning, bypass pipe (1) allows the cold water to enter the conventional water heater directly.

through the heat exchanger. When water is hard, containing dissolved minerals, especially calcium, the high temperatures in the heat exchanger tend to force the minerals out of solution and a hard mineral deposit is formed inside the pipes or tank of the heat exchanger. Occasional cleaning or replacement of the heat exchanger is necessary.

A second problem can be corrosion inside the system, particularly if it is made of iron or steel (not stainless). Certain dissolved materials in natural water accelerate this corrosion, and are taken out of the water in the process.

Water recirculated again and again through the heat exchanger soon loses its dissolved calcium and corroding ingredients and is then quite harmless to the heat exchanger. Such a recirculating system requires a second heat exchanger to get the heat into the domestic water, as illustrated in Figure 5–16. Another possible advantage of this system is that the recirculating loop need not be pressurized (or "closed" in the sense that makes

explosions possible). Thus less expensive materials for tanks and piping can be used, and an explicit expansion tank may not be necessary.

6. *Prevention of storage-tank bypassing.* In some kinds of installations (Figure 5–17) there is a tendency for only warm or even cold water to come out the hot water faucets even when the storage tank is hot. This happens if the "hot" water is drawn directly from the heat exchanger rather than from the tank. As discussed previously, one quick pass through most heat exchangers does not warm the water very much. If this bypassing of the storage tank is total, the system is virtually useless as a water heater. If the bypass is partial, with the "hot" tap water coming partly from the heat exchange and partly from the storage tank, the system can work, but at reduced output and efficiency.

Bypassing can only occur if a common tap on the storage tank is used both to feed hot water from

115

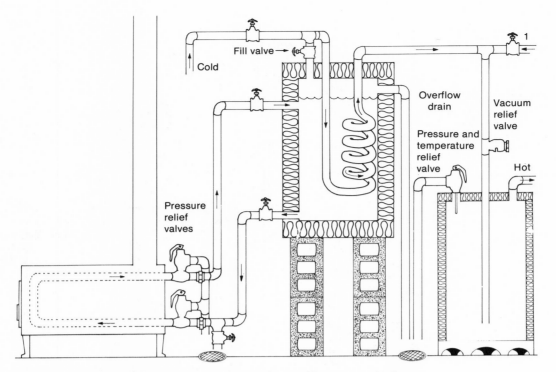

Figure 5–16. A recirculating or closed loop wood water heater. The same water keeps recirculating through the wood stove; this minimizes corrosion and hard water deposits in the stove coil. This necessitates a second heat exchanger (the long helical coil in the insulated storage tank) to get the heat into the tap water. Concentric tanks may also be used. Make-up water must be added occasionally, particularly if the system is open; natural circulation will not occur if the water level falls below the hot water input to the tank from the stove. Since keeping the water circulating is so important, use of a float valve to add make-up water automatically is advisable. Alternative cold water input (1) is used when wood system is not in use.

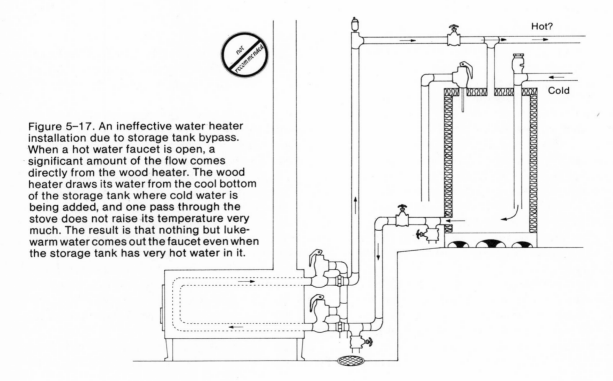

Figure 5–17. An ineffective water heater installation due to storage tank bypass. When a hot water faucet is open, a significant amount of the flow comes directly from the wood heater. The wood heater draws its water from the cool bottom of the storage tank where cold water is being added, and one pass through the stove does not raise its temperature very much. The result is that nothing but luke-warm water comes out the faucet even when the storage tank has very hot water in it.

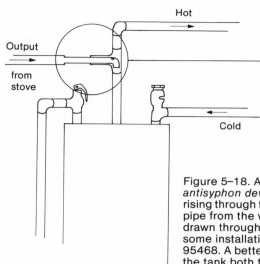
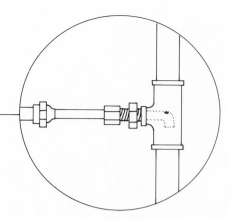

Figure 5–18. An anti-storage-tank-bypass device, also called an *antisyphon device*. When a hot water faucet is open, the hot water rising through the T from the storage tank creates back pressure in the pipe from the wood water heater preventing much water from being drawn through this pipe. One source for such a device, suitable in some installations, is Blazing Showers, Box 327, Point Arena, CA 95468. A better solution when possible is not to use the same tap on the tank both to get wood-heated water in and to get hot water out.

the wood heater into the storage tank, and to pass hot water from the storage tank into the hot water pipes of the house, as illustrated in Figure 5–17.

One way to avoid this problem is to use the pressure and temperature relief valve (PTRV) tap for the incoming water from the wood heater. A PTRV must still be used at this tap. Since its new mounting will be a little higher up (above an added T), a new PTRV may be necessary to assure the temperature-sensitive stem is still properly immersed in the storage tank water. This solution to the bypassing problem has been incorporated, if needed, in most of the systems illustrated in this book. This solution has a potential drawback; the PTRV may open frequently if the water coming from the wood heater is very hot. In systems with adequate circulation and proper sizing, this should not occur.

Another solution is use of so-called "anti-syphoning" devices (Figure 5–18). These prevent water from being drawn through the heat exchanger when a hot water faucet is open.

Safety of Boilers and Water Heaters

Explosions are the greatest danger in *any* system that heats water in closed containers such as domestic water heaters, central heating system boilers, fireplace and cook stove water heating accessories, or pressure cookers. Such explosions are sometimes called BLEVE's, an acronym for Boiling Liquid Expanding Vapor Explosion. When water is heated, it expands slightly, but with incredible force. A very large expansion occurs if it is allowed to boil. Whether or not boiling occurs, the pressure continues to rise in a closed system of water that is being heated. At some point the strength of the container may no longer be a match for the expansive forces of the water or steam, and the container can burst. This can be at a temperature below the normal boiling point of water, but normally occurs at higher temperatures.

The results can be catastrophic. Steam and hot water burns are possible. Superheated water emerging from a pressurized boiler flashes into steam with explosive suddenness and force. But usually the metal projectiles are even more dangerous. In some cases whole conventional water heaters minus their bottoms have shot like rockets through two floors and a roof! Cast-iron water jackets in wood stoves can explode into pieces of shrapnel with fatal consequences (Figure 5–19). Steel and copper containers and pipes often "tear" without releasing as many pieces of metal but the possible consequences are still very serious.

Prevention of steam and hot water explosions of water heaters and boilers requires careful attention to the following four points:

1. *Use of adequately strong and durable materials.* Tanks and pipes must be designed to withstand the highest expected pressures with a large

Figure 5–19. Why automatic pressure relief valves are necessary in all water-heating systems. Shown are the remains of an antique cook stove. It had a cast-iron water-heating tank that was capped shut with some water in it. The explosion of the tank destroyed the stove and resulted in the loss of a leg of one person. (Photograph contributed anonymously.)

margin of safety and to have reasonable durability. Boilers and all "pressure vessels" must generally be approved by the American Society of Mechanical Engineers (ASME) in order to be permitted under most building codes, although in some states small systems are excluded. For homemade systems, steel or brass pipe of types used in domestic plumbing are reasonable for use as a heating coil in a stove or fireplace, but the threaded joints should be welded or brazed. Copper tube and pipe are often used but are more easily damaged by the impact of logs and deteriorate more rapidly at high temperatures. If the system is never fired without water in the copper heat exchanger, no high temperature deterioration can occur; even ordinary solder can be used, although silver solder is preferable. Copper is always suitable for use *outside* the combustion chamber or chimney, such as in a hot air plenum or between walls of a circulating stove.

2. *An expansion tank.* This is necessary in most *closed* systems such as typical hot water boiler central heating systems, and in rural well water systems. Within normal temperature ranges encountered in boilers (roughly 55°–180° F.), the volume of the water in the system will change by about 3 percent. Under extreme conditions, the water temperature may range from 40° to 250° F., with a corresponding volume change of 6.1 percent. With no room inside the system for this expansion, the pressure relief valve would frequently open, which would result in a wet floor, wasted water, and unnecessary corrosion or calcification inside the system. And if the pressure relief valve should fail, the system would explode. Closed hot water heating systems are always installed with an expansion tank to absorb the volume changes of the water, and the tank's capacity is typically around 10 percent of the water capacity in the whole system. Proper sizing of expansion tanks is complicated and should be figured by a competent professional. But note that adding a hot water storage tank or a supplemental boiler to an existing system usually necessitates a larger or an added expansion tank.

3. *An automatic pressure-relief system.* Common systems consist of a valve (Figure 5–20) that opens automatically when the pressure exceeds a certain limit, and a pipe that directs the vented water or steam towards the floor or other safe and relatively convenient place.

Pressure relief valves are rated both in terms of the pressure at which the valve will open, and in terms of power—Btu per hour or watts—which is the rate at which heat energy can be expelled through the valve when it is open. Combined pressure and temperature relief valves are normally used on water heaters.

The pressure rating of pressure relief valves should be well above the normal operating pressure inside the system, and well below the pressure/strength limitations of the system's tank and piping. A rating of 150 pounds per square inch (psi) is typical for values on a conventional water heater; 30 psi is typical for many hot water boilers. For an essentially open or vented system such as the recirculating loop in Figure 5–16, 30 psi is also reasonable if the vertical extent of the system is no more than about 40 feet. Pressure relief valves *are* needed even in open systems if there are isolation valves, which there normally are.

The power capacity of a pressure relief valve must be adequate to handle the maximum conceivable power input to the water from the heat source, and a large margin of safety is wise. For

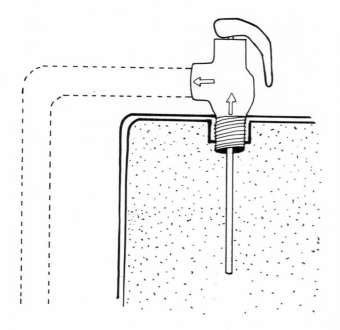

Figure 5–20. A pressure and temperature relief valve (PTRV) shown schematically installed on a hot water heater. The stem extending down from the valve is the temperature sensor and must be immersed in the hot water. Either high pressures or high temperatures inside the water heater force the valve open allowing steam or hot water to be vented out. The discharge pipe must have no size reduction and should extend to 6 to 12 inches above the floor, to prevent steam burns to any person standing nearby. If the valve blows very often, it will be convenient to have the discharge pipe over a floor drain.

ordinary water heaters, 150,000 Btu per hour is typical; for ordinary boilers, 300,000 Btu per hour is typical. For most solid-fuel systems the same ratings are reasonable.

It is wise to install *two* pressure relief valves in direct-fired coil- or pipe-type water heaters, one at the input and one at the output end of the coils. Hard, rock-like lime deposits inside the coil from minerals in hard water can so restrict the coil that expansions cannot relieve through the constriction to a relief valve on the other side.

Lime deposits can be removed with lime solvents available at many hardware and plumbing supply stores. Two manufacturers are Nyco Products, 1801 South Jefferson, Chicago, Ill., and S.O.S. Products, 34 Cumberland St., Brooklyn, N.Y. Instructions with the solvent should be followed. The procedure involves draining the water out, pouring the solvent in, and then draining and flushing the solvent out.

A pressure relief valve must be installed with no valve between it and the heat source. This is to prevent inadvertent isolation of the relief valve

from the expanding water. The valves are generally located no further than 6 inches from the tank or coil needing protection. The discharge from the valve should be piped without pipe-size reduction to about 5 inches above the floor or drain.

Adequate pressure relief in water heating systems is *not* given through the cold water main of a town water system. Plumbing the wood water heater into a conventional water heater that has a pressure relief valve of its own offers some protection but is also generally considered inadequate. The principal reason is that every water heater must have pressure relief that cannot be separated from the heater by closing a valve. Since virtually all installations have isolation and shut-off valves for maintenance purposes, there must be a pressure relief valve *at the water heater*. A second, less important reason is that it takes time for pressure surges at a heater to be transmitted to a distant relief valve. If a sudden pressure surge were to occur, a remote pressure relief valve would not necessarily give the needed protection.

Pressure relief valves are so important for safety that they should be tested at least once a year. Hard water deposits can seal them shut. The lever on the top of the valve should be pulled up to assure the valve is free. Water or steam should come out of the discharge pipe. Unfortunately after being opened, the valves do not always seal again—particles can get lodged in the seat. The valve may then need replacing to avoid constant dripping and water loss. (Some valves do not have this testing lever. I recommend against using such valves.)

4. *Prevention of a dry (empty) system from being fired.* Without water in them, water heaters and boilers can become much hotter than normal. This can be damaging to the heaters themselves, but perhaps even more serious would be the possible steam explosion if water were to enter such an overheated system. When water contacts metal that is considerably hotter than the water's boiling point, voluminous steam is instantly generated. The resulting expansion can be far greater than even a pressure relief valve can handle; even systems with pressure relief valves can explode. This is most likely to occur in homemade water-heating and hydronic house-heating systems. In manufactured conventional systems this type of accident is rare; automatic addition of make-up water tends to keep the units full, and if loss of water does occur, a professional is likely to be overseeing the refilling. Many fossil-fuel steam

systems have a low-water cutoff that does not permit the gas or oil burner to operate if the water level is low.

With pipes or tanks in fireplaces and stoves, the danger is very real. Gravity-flow systems, where the heat exchanger is at the lowest point in the system, are relatively safe since the entire system must go dry before the heat exchanger could be dry. Pumped systems where the heat exchanger is not at the lowest point in the system are vulnerable, particularly if they drain down when the pump is off, as in some solar systems, or if fired after the system has been filled with water but before the pump has been turned on, or at any other time when air may be in the heat exchanger. Use of a water level sensor is illustrated in Figure 5–14.

The steam explosion hazard is very real. Good engineering is essential to minimize it. A second, less serious safety hazard associated only with water heating is painfully hot water or steam reaching hot water faucets. The results can be painful and messy, although serious injury is unlikely.

The heat source in ordinary water heaters is controlled by an *aquastat* (thermostat that senses water temperature). The aquastat shuts off the burner or electricity when the water reaches a preset temperature, less than boiling, typically between 140° and 180° F.

Wood boilers with domestic hot water capabilities can produce overheated tap water. Although most boilers have aquastats to regulate boiler water temperature by controlling combustion air, overheated water is still a common occurrence due to operator negligence or even intent, or due to leaked combustion air when the automatic damper is shut. Thus the probability of overheated tap water is generally higher in wood-fired systems than in conventional systems.

Two types of temperature-limiting devices are suitable for wood water-heating systems, and both should be used. One is the temperature part of a temperature and pressure relief valve; the valve is opened either by excessive pressure or by temperatures in excess of 210° F. For temperature control purposes the valve may be located anywhere on the hot water side of the system, most often at the storage tank, and usually is combined with a pressure relief valve. Locating a temperature relief valve too close to a radiant stove may cause the valve to open annoyingly often.

The other temperature-limiting device is a

tempering or mixing valve. A temperature-sensitive element inside the valve causes cold water to be mixed with the hot to control the output temperature. The valves are usually adjustable over a range of output temperatures. They are commonly used with tankless coil heaters in oil and gas boilers. Although they are not critical for safety, I recommend mixing valves. Storage tank temperatures often fluctuate widely—mixing valves help steady the temperature of the delivered water. The valves also effectively extend the storage capacity of the system by preventing full flow from the storage tank when its water is hotter than needed.

Heat Extractors

Heat extractors, also called *heat reclaimers*, *heat robbers*, and *heat savers*, are accessories designed to extract heat from the flue gases (Figure 5–21). They can be useful and economic accessories in systems where heat transfer efficiency is low, but are not needed in high-efficiency systems.

Heat extractors should be easy to clean without disassembly. The better they are at extracting heat, the cooler are the inner heat transfer surfaces, and the more creosote builds up there, ultimately preventing efficient heat transfer. Daily cleaning is often needed to maintain efficiency. One convenient cleaning method provided on many tube extractors is a plate inside the device with holes that just fit around each tube. Using an attached rod and handle, one can slide the plate back and forth once over the tubes scraping them clean. The creosote falls down into the stove. The operation takes only a few seconds.

Heat extractors take heat from the flue gases, and this leads to less draft and more creosote accumulation. Thus I do not recommend adding a heat extractor to a system with marginal draft, or to a system with creosote problems unless more frequent chimney cleaning is accepted as required additional maintenance. On the other hand, in systems with relatively clean-burning fires or very hot flue temperatures, heat extractors may not result in much more creosote accumulation.

To help minimize creosote, heat extractors with thermostatically controlled fans should be designed so that the fan only comes on when flue gas temperatures exceed some high temperature, perhaps around 400° F. This is a guess, not a verified reasonable compromise between heat extraction and creosote minimization. (See

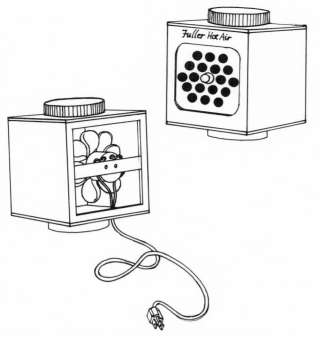

Figure 5–21. A tube and blower heat extractor or reclaimer.

"Monitoring Smoke Temperatures" in Chapter 3.) Four hundred degrees Fahrenheit *is* the temperature specified in the Underwriters Laboratories (UL) standard for heat reclaimers.

Heat extractors are capable of burning out, just as are many other components in a wood heating system. The inner portions of the tubes in the tube-type extractors are especially susceptible because of the higher temperatures there. Use of heavy gauge material or stainless steel will prolong the useful life of the device. But no kind of practical construction is immune from burnout. Even UL-listed heat reclaimers can burn out. Thus, just as with stovepipe, periodic inspection is necessary, with repair or replacement as needed. Mechanical weakness of the tubes can be discovered prior to actual burnout by poking the walls with a metal rod or long screwdriver.

As is the case with most electrically powered wood-heating equipment, the most severe conditions usually occur during a power failure. The tubes in heat extractors can become extremely hot if the fan or blower is not operating. When the tubes burn out, the fumes from the fire can enter the house directly. When power is restored, the blower may blow flames and sparks out onto the floor. Very hot fires should be avoided whenever the heat extractor's blower is not operating.

Use of Coal

Use of coal as a fuel in residential heaters has essentially all the same general types of safety considerations as does use of wood. The same types of chimneys, the same installation clearances and floor and wall protectors are appropriate for both coal and wood appliances. The chemical composition and physical form of coal creosote is slightly different, but the danger of chimney fires is similar, as are the techniques for prevention—frequent inspection and cleaning, and operating the heater with small hot fires rather than large smoking fires. Back flashing, chimney flow reversal, and careless ash disposal are common concerns with all solid fuels. The practical aspects of lighting and maintaining fires are somewhat different with coal.

Coal should not be burned in a device designed only for wood fuel, but wood may be used safely in a coal-burner. Coal fires tend to be hotter than wood fires and thus may damage a wood-only heater. Coal heaters can usually be identified by their grates. Coal requires grates with relatively large holes or gaps because of the large size of coal ash particles called clinkers. Coal grates are usually designed to be shaken or rotated to help the clinkers fall through. Because of the higher fire temperatures, coal grates are usually more massive than wood fire grates. Most contemporary coal heaters have firebrick liners.

Use of Other Special Fuels

Other than wood and coal, there are three other forms of fuel sometimes used in "wood" burners—newspaper logs, wax-sawdust logs, and densified wood. Although available evidence is meager, two of these fuels may be hazardous in some applications.

Wax-sawdust logs are made of sawdust or small wood chips held together with paraffin and formed in the shape of round logs. Instructions usually say to use them only in fireplaces, only one at a time and, not on top of a bed of hot coals. They should not be poked or stirred when they are burning. The wax component is what can lead to difficulty. Wax is a very concentrated or dense fuel. One whole wax log burning naturally by itself does not constitute a hazard. But more than one log or a broken burning log can result in a very hot and potentially dangerous fire. There have also been reports that burning more than one log at a time or using the logs in some kinds of stoves can result in melted wax flowing out of the combustion area. Since this wax may be burning, the consequences can be serious. But when used as directed, wax-sawdust logs appear to be safe.

Densified wood is highly compressed sawdust, chips, and other organic waste. It has little or no added substances as a binder. The compression results in a product roughly two to three times denser than natural wood, so dense that it *sinks* in water. It is produced as solid cylinders with diameters ranging from a fraction of an inch up to about four inches, depending on the intended use.

The possible hazard associated with use of densified wood is too hot a fire. Denser fuels tend to burn hotter. Densified wood is similar to coal in a number of respects. Thus, until more experience is gained, I would suggest treating it like coal by not burning it, at least in large quantities, in wood-only burners.

Densified wood is also usually very dry. As discussed in Chapter 3, this dryness may result in considerable creosote when the wood is burned in a relatively airtight appliance. The dryness and relatively small piece size both may also increase the possibility of flashbacks. Densified wood has great potential as a fuel for automatically fed central heaters, but I recommend caution (small loadings) when using it in most currently available equipment.

Use of newspaper logs also needs more research to establish whether there are any hazards. I am aware of none. As a practical matter, it is sometimes difficult to sustain a fire with newspaper logs alone. Glossy paper, the type used in many magazines, has a very high ash content and is thus more difficult to burn.

Burning household trash carelessly can result in dangerously hot fires. Corrosion of the stove and steel stovepipe may also be a problem; the burning of some types of plastics generates corrosive gases. But burning *small* quantities of *plastic-free* combustible trash is not hazardous. Use of salt-water driftwood also tends to be corrosive to wood heating systems.

Mobile Home Installations

Fundamentally, heating with wood in a mobile home does not result in hazards not encountered in an ordinary home. However some safety issues are more critical. The small size of most mobile homes increases the likelihood that outside combustion air will be necessary (see discussion of outside air later in this chapter). Mobile home fires tend to spread more rapidly because of the use of more flammable construction materials. And the higher proportion of synthetic materials results in a more lethal smoke.

To be legal, mobile home installations must often satisfy some different conditions because mobile home safety is under a different jurisdiction, the federal Department of Housing and Urban Development (HUD). Some particular areas where HUD requirements for mobile home wood stoves differ from general code requirements are:

1. Stoves may not be installed in bedrooms.

2. Stoves must be equipped with the means to be secured to the mobile home floor, so that they can be made to stay in place if the home is moved.

3. The chimney system must be listed factory-built.

4. The chimney must attach directly to the wood heater. *No* stovepipe connector is permitted. (In some cases special kinds of "stovepipe" are included as part of the stove and are thus legal.)

5. A spark arrester and cap must be used at the chimney top.

6. The heater must obtain its combustion air from outdoors via a duct that feeds directly into the combustion chamber. The duct must be designed so that it is impossible for hot coals to drop down it to the ground under the mobile home, and a rodent guard (screen) must be used. The duct area should be at least half the flue area.

7. The installation should not result in more than 40 pounds per square foot of loading on the floor.

Mobile homes tend to be constructed with less margin in terms of structural strength. Extra care must be taken not to weaken the structure when installing a chimney. Ideally, *no* structural member should be cut. In practice, if a stud or joist *must* be cut, some reframing and reinforcing is advisable.

Fireplaces

Many of the safety considerations of heating with wood are common to all types of wood burners. But some hazards and precautions are particularly or only relevant to fireplaces, and in some instances to fireplace stoves.

Controlling Sparks

Since fireplaces are usually intended to be operated as *open* wood burners (some, of course, have the option of closed-door operation), containing the exposed fire is important. Andirons are needed to keep logs from rolling out of the fireplace. Chasing burning, rolling logs across the living room floor may be a challenging sport, but I would recommend frisbee as a safer alternative. Spark screens or closed doors should be in place most of the time a fire is burning to prevent sparks from starting a house fire. I was interested to discover in some experiments that the wire-cage spark screen I tested only reduced fireplace energy efficiency by one or two percentage points.

An adequately large hearth extension in front of the fireplace will minimize the effects of sparks or logs which *do* get out of the fireplace. Hearth extensions are the non-combustible floor coverings in front and to the side of fireplace openings. Bricks, slate, and tiles are common materials. NFPA recommends hearth extensions 16 or 20 inches beyond the fireplace opening for masonry fireplaces depending on whether the fireplace opening is less than or more than 6 square feet in area (Figure 5–22).

For people who make heavy use of an open fireplace without a spark screen, I suggest a considerably larger hearth extension. Many codes require floor protection 36 inches beyond the front of certain types of radiant gas heaters, and some prefabricated fireplace manufacturers are finding that 16 to 20 inches is not always enough to prevent floor overheating from fireplace radiation alone. Thus for a heavily used unscreened open fireplace, I recommend floor protection of 24 to 30 inches beyond the fireplace opening. Small dark charred spots on floors or rugs around existing fireplaces are a sign that the floor protection does not extend far enough. Appropriate protecting materials are those discussed in Chapter 2 for protection under stoves.

A danger with any size add-on hearth extension that is not a part of the fireplace itself is that sparks and hot coals may slip down between the fireplace and the extension. Mortar and/or sheet metal should be used to assure a spark-proof joint between the fireplace and its hearth extension.

When installing a metal fireplace, a possible extra precaution is to assure there is no insulation in the floor under the unit. All listed wood heating systems are tested when installed on an uninsulated floor. This permits considerable heat to dissipate down through the floor. If the floor is insulated, it may overheat. More research is needed to determine if this is a serious problem requiring changes in the safety testing of such appliances. But until this is done I recommend removing any insulation in floor directly under a metal fireplace, particularly zero-clearance fireplaces, unless a ventilated floor protector is used (Figure 2–60).

Chimney Temperatures and Clearances

Heavy use of some fireplace accessories and modifications can result in dangerously high

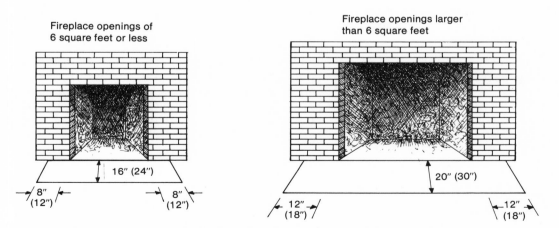

Fireplace openings of 6 square feet or less

16" (24")

8" (12")   8" (12")

Fireplace openings larger than 6 square feet

20" (30")

12" (18")   12" (18")

Figure 5–22. NFPA recommended minimum dimensions for hearth extensions in front of fireplaces. A number of authorities, including the author, feel these dimensions should perhaps be larger, by about 50 percent, particularly for heavily used fireplaces. These larger dimensions are indicated in parentheses in the figure.

chimney temperatures. The flue gas of an ordinary open fireplace is rarely very hot because of the considerable amount of room air entering the fireplace and diluting the hot smoke. Accessories or additions that still allow a large fire but that restrict this excess air will usually increase chimney temperatures. Examples are glass or metal doors with or without heat extraction equipment such as tubular grates, and large attached stoves or fireplace stoves. Using such equipment with large fires continuously for many hours can substantially increase the risk of a house fire unless the fireplace and chimney have proper clearances to combustible materials.

Existing and proposed standards for masonry chimneys are discussed in Chapter 2. I believe that there shouldn't be contact between wood or other combustibles and the chimney. Few installations conform to this standard or even the less stringent NFPA recommendations. Thus I would be cautious about having intense fires lasting more than three hours in masonry fireplaces with closed doors. As always, smoke detectors, including one in the attic if applicable, can significantly reduce the possible consequences of an overheated fireplace and chimney.

Most building codes state that no other appliance should share the same flue passage with an *operational* fireplace, as explained in Chapter 2. However, if another appliance is connected to a fireplace flue, keeping a spark screen or doors in place at all times, even when the fireplace is not in use, and keeping the damper closed whenever possible will minimize the danger. These measures should be taken anyway to prevent the fireplace's own sparks from getting out and to prevent unnecessary loss of heated indoor air up the chimney when the fireplace is not in use.

Circulating Fireplaces and Fireplace Furnaces

A special concern with factory-built circulating fireplaces is the potentially high temperature of the heated air. High-temperature air in ducts can ignite nearby combustible materials such as wood. Most codes do not explicitly address this issue, but a circulating fireplace whose hot air is ducted away, if only for a few feet, is essentially a wood furnace. Hand-fired solid-fuel furnaces require 18 inches of clearance to combustibles from the plenum and the first 3 feet of ducting, 6 inches for the next 3 feet and 1 inch for much of the rest. For a circulating fireplace with hot-air ducts, these

NFPA furnace clearances would seem to translate to at least 18 inches for the first 3 feet of duct, at least 6 inches for the next 3 feet, and at least 1 inch thereafter.

These clearances are designed for worst conditions—continuous and intense firing of the appliance. Although you may not intend to use a fireplace this way, an oil shortage or power outage can change your plans. During a power failure, any blowers in a fireplace will stop extracting heat. This will result in even higher air temperatures. Since such large clearances are rarely practical, protection of the combustibles (Table 5–3) or use of insulated or double-walled ducts is usually required. Clearances may be as little as 1 inch if UL-listed double-walled metal chimneys such as Type L, Type B, and Type B-W vents are used as ducts.

Many new prefabricated circulating fireplaces are "listed" appliances. This means their detailed installation instructions have been checked for safety by an independent laboratory. All clearances required for safety will be explicit in the instructions, and these clearances supersede any NFPA recommendations. All instructions must be followed carefully. It is dangerous to follow one's intuition here.

Glass Doors

Some fireplace accessories can be, or can become, hazardous. Large glass doors are a relatively new development. Considerable breakage has occurred, but breakage is usually an annoyance, not a serious safety hazard. Tempered Pyrex® glass is the choice of many manufacturers. The tempering adds strength but is gradually lost at high temperatures, and causes the glass to fragment and fly apart when broken. When the door is properly designed and used, tempered Pyrex seems to be a satisfactory material. Alternative kinds of glass such as Vycor® are far superior to tempered Pyrex® in terms of thermal shock resistance, but are more expensive. Vycor's impact strength is also less for equal thicknesses.

Glass doors on ordinary fireplaces with no other accessories apparently *decrease* the energy efficiency of the fireplace. (See paper by the author listed in the Bibliography.)

Tubular Grates

Tubular grates and other air-circulating fireplace accessories can burn out. This is not likely in the better quality tubular grates, which are made out of stainless steel. If holes develop, usually in lower portions of the grate, smoke and carbon monoxide can be pulled into the moving air stream inside the pipe and end up in the house. The odor should be an adequate warning that something is wrong. Smoke detectors are always good insurance. Particularly in grate units with blowers, sparks can also get picked up in the air stream and spewed out into the room. Thus, as always, periodic inspection of equipment is wise.

Chimney-Top Dampers

There is an unusual kind of chimney damper available for masonry chimneys serving fireplaces. The damper is at the top of the chimney and the damper's position is controlled from the fireplace inside the house. Such dampers have the potential advantage of preventing cold outdoor air from descending into and cooling the chimney and then the house when the fireplace is not in use. It is absolutely critical that the damper never shut due to heat, breakage, or wind, when there is a fire in the fireplace. As a practical matter, such dampers should also not be damaged by chimney fires. The dampers should probably also be designed so they cannot freeze shut; people often light fires before remembering to open the damper. Until both the safety of such devices is clear and the possible beneficial effects are quantified, I would be hesitant about using them.

Fireplace Boilers and Water Heaters

Fireplace tap-water or hydronic heating systems can explode just as can any water-heating system. All the safety precautions discussed earlier in the chapter should be observed and a competent plumber should be consulted.

Outside Air

Ducting outdoor air directly to a wood-burning appliance is not necessarily beneficial, but in some cases is almost essential. It all depends on the objectives and on the particular installation.

One myth about outside air is that without it, there is danger of oxygen depletion of the air in the house. Wood fires cannot selectively extract oxygen from the air in a house without simultaneously emitting smoke into the house, and the smoke is probably more hazardous than the

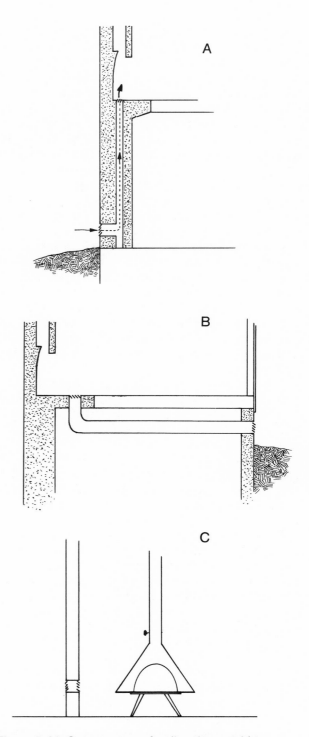

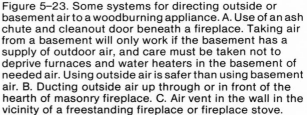

Figure 5–23. Some systems for directing outside or basement air to a woodburning appliance. A. Use of an ash chute and cleanout door beneath a fireplace. Taking air from a basement will only work if the basement has a supply of outdoor air, and care must be taken not to deprive furnaces and water heaters in the basement of needed air. Using outside air is safer than using basement air. B. Ducting outside air up through or in front of the hearth of masonry fireplace. C. Air vent in the wall in the vicinity of a freestanding fireplace or fireplace stove.

slightly lowered oxygen levels. Fireplaces and stoves take in *total* air, which is roughly 20 percent oxygen and 79 percent nitrogen, plus very small amounts of other gases. Only if the combustion products are vented back into the house is oxygen depletion possible for only then is the oxygen-depleted combustion air returned to the house. This occurs intentionally with some types of natural gas- and kerosene-fueled room heaters and with most gas ranges. It is never intentional with wood heaters because the smoke would be unbearable. Even if oxygen does become somewhat depleted, it is not the decrease in oxygen levels that is most noticeable, but the presence of carbon monoxide and all the other ingredients of wood smoke.

Outside air is needed in houses that are so tight that either the appliance does not work properly or use of the appliance interferes with the use of other appliances in the house. Modern energy-conserving homes, some mobile homes, electrically heated homes, and earth homes tend to be tight.

Tight houses do not generally impair the operation of wood stoves because stoves use so little air (10 to 50 cubic feet per minute). But even if a house were so tight as to restrict the air supply available to a stove, this is rarely hazardous. Performance, not safety, suffers. With a decreased air supply a stove's fire cannot be as hot and so the heat output of the appliance is decreased. But most houses have enough air leakage to supply stoves adequately.

Fireplaces and fireplace stoves often will not work satisfactorily in a tight house. *Open* wood burners require a certain minimum average velocity (about 0.8 feet per second[1]) of air into the fireplace opening to prevent smoke from coming out the opening into the room. This amounts to between 50 and 1500 cubic feet of air per minute depending mostly on the size of the fireplace opening. If a tight house cannot supply this amount of air, smoke will spill out into the room from the fireplace. Thus the appliance may be unusable unless a window is opened or an outside air system is installed.

Another clearly justifiable reason for direct outside air is interference with other appliances. Use of a fireplace tends to depressurize the house through the draft or suction of the fireplace chimney. Exhaust and attic fans can have the

same effect. This can prevent other vented appliances such as furnaces, boilers, water heaters, and even other fireplaces or stoves from venting properly, resulting in carbon monoxide an other lethal gases accumulating in the house. It is not always easy to anticipate if this will be a problem in any given situation. If, when the fireplace is in use and the other appliances are not, cold air is moving down the other chimneys into the house, there is potential for this kind of interference. The solution is to supply outdoor air either to the fireplace or to the other appliances.

A rough estimate for the appropriate size of an outside air duct system is half the cross-sectional area of the appliance's chimney flue.

Another benefit of ducted air is decreased drafts at floor level in front of a fireplace and thus increased human comfort.

Use of ducted outside air can also decrease the chances of chimney flow reversal and resulting wood-smoke asphyxiation. A smoke detector is the best protector. Whether a duct is attached directly into the appliance or there is just a fresh-air register in the vicinity, a pressure-equalizing link to the outdoors on the ground floor or basement of a house will usually prevent flow reversal and ensure that the chimney will be self-starting. An outside-air hole through an upper-story wall of a multi-story house will be counter-productive and should not be installed.

Perhaps the most common reason given for using ducted outdoor air is a supposed improvement in energy efficiency through elimination of the flow of warm house air up the chimney. The reasoning is most plausible in the case of fireplaces because of their air-guzzling tendencies.

I feel that the claims for ducted outdoor air saving energy need more substantiation. Although it is clearly beneficial to decrease excessive house-air losses, some other counter-effects in some cases may be important, resulting in a negative net energy impact—more oil, gas, electricity, or wood consumption because of the outdoor air systems.

Here are some possible negative effects:

1. Use of colder outside air decreases heat transfer efficiency, because the fire and all surrounding heat transfer surfaces are cooler. Accurate quantitative predictions are difficult, especially for fireplaces, but for stoves I estimate that the decrease in heat transfer due to use of cold outside air is roughly the same as the heat saved by

1. *1975 Equipment Volume, ASHRAE Handbook and Product Directory* (American Society of Heating, Venting and Air-Conditioning Engineers, New York, 1975) p. 26.26.

not warming the combustion air up to room temperature; thus it may be a break-even proposition. But such predictions are highly uncertain; measurements are needed. However if it turns out that the heat saved is offset by the decrease in heat output, then the overall net effect is likely to be negative because of the following mechanism.

2. Almost inevitably, having a duct for outside air to an appliance in the living space of the house will result in cold air leakage into the house whether or not the appliance is in use. Ducts and dampers are rarely airtight. Even if there is no leakage directly into the house but only into the appliance, the appliance and its chimney will be cooler when not in use and more heat will conduct out of the living space of the house into the appliance and its chimney. This kind of heat loss due to an installation itself and when the appliance is not in use might be termed standby losses, and these losses can be larger for systems with direct outside air.

3. Combustion efficiencies may be lowered due to lower temperatures in the combustion region.

4. If the tightness of the house is a limiting factor in the amount of air entering a wood burner, supplying direct outside air may increase the total amount of air entering the appliance. In many cases, particularly fireplaces, this will increase the amount of air not contributing to combustion and this in turn will decrease heat transfer efficiency. I expect this effect is most important for circulating-type open fireplaces.

Only in the case of closed wood burners such as stoves, or fireplaces with tight-fitting closed doors is it clear that most of the room air loss can be eliminated. An outdoor air register in front of an open fireplace is unlikely to eliminate all room air from entering the fireplace opening. But in the case of closed wood heaters, the need to eliminate room air loss is less because closed burners with tightly fitting doors do not consume all that much air anyway. Thus to some extent outside air systems may be most effective in cases where they are least needed. But again, what is really needed to resolve the issues is good experimental measurements of the net effect in real-life situations.

My personal conclusions on use of ducted outdoor air are:

● It can be a necessity for open fireplaces and fireplace stoves in very tight houses unless one is willing to have a window open when using the fireplace.

● It can prevent improper and dangerous spillage of fumes from *other* appliances into the house.

● It can increase floor-level comfort in front of the wood-burning appliance.

● It can significantly decrease chances of chimney flow reversal and can eliminate the non-self-starting property of some chimneys.

● There is a reasonable doubt about its effect on net energy efficiency, which may well be negative. If further research proves this to be the case, I would not recommend direct outdoor air except in cases where there is a real need for one of the benefits listed above.

There are two very important safety considerations in designing outside air systems. One concerns the danger of the air input system being used as a chimney, with hot flames or gases getting into the air duct. If the air system does not terminate inside or on the appliance, but ends in the general vicinity of the stove, fireplace, or furnace, there is little hazard. But if the air system connects directly to the appliance, some precautions should be taken. It is best to locate the outdoor air intake lower in elevation than the appliance, and on the side of the house that receives the prevailing winds. These measures tend to increase the air pressure in the duct, which encourages flow in the proper direction. If possible, the ducting nearest the appliance should slope upwards towards the appliance, even if elsewhere the ducting slopes the other way. This helps turn around a momentary flow reversal such as may be induced in gusty weather. The British suggest using two ducts from two walls at right angles to each other, with the ducts meeting in a "balancing" chamber before the final connection to the appliance.[1] This is intended to alleviate windy weather problems.

In rare cases of buildings with positive pressure inside relative to outside, no type of simple passive outside combustion air system will work at all unless the appliance is very tight and is only

---

1. "Code of Practice for Installation of Domestic Heating and Cooling Appliances Burning Solid Fuel," CP 403 (London: British Standards Institute, 1974), pp. 14, 15.

operated with a closed combustion chamber. Examples are buildings with positive-pressure air make-up heating and ventilating systems.

The other hazard of outside air systems is that of coals or sparks falling into the air duct system. Where this is possible, such as in ducts connected directly to appliances, or masonry fireplace air systems in the fireplace hearth, one should use non-combustible materials and have an inch or two of clearance between the duct and combustibles wherever the hot coals could come to rest in the duct.

Epilogue

In this book I have tried to describe and explain all the major hazards of heating with wood and coal, and to describe and explain all the precautions that can be taken to avoid these hazards.

I am concerned that very conscientious people may try to take every conceivable precaution. This would be excessive since many precautions have overlapping functions. For instance, there are many ways to mitigate the dangers of chimney fires. One is through installation of a very expensive chimney which is essentially chimney-fire safe. Another is to inspect and clean the chimney as often as necessary so the buildup could never fuel a big chimney fire. Another is to design and operate the system so that essentially no creosote accumulates. There is no denying in this example that taking all precautions yields the safest system. But each added precaution brings diminishing returns. Safety can also be excessive in the sense of detracting from the usefulness and efficiency of the system. An example would be installations which send most of the fire's heat up the chimney to keep the chimney clear of creosote.

My hope is that I have made these and other compromises and trade-offs sufficiently clear so that the reader can recognize a reasonable, adequately safe, and moderately efficient wood heating system. It may be a difficult fact to accept, but there is no *best* system when all things are considered.

I am also concerned that I may have scared some people away from wood heating. I hope not, because these are the very people who, if they did heat with wood, would do it most safely. A healthy respect for fire is perhaps the most important factor in heating with wood safely. And respect in this case is fear tempered with knowledge. If you have read this book you have the knowledge. So go heat with wood and enjoy it.

Mechanisms for Spontaneous Combustion, and "Safe" Temperatures for Combustible Materials

Many house fires started by wood and coal heating systems begin through spontaneous ignition of wood near a hot stove or chimney. Installation clearances and use of protective panels are designed to prevent this by keeping nearby combustible materials well below dangerous temperatures.

Webster's *Third New International Dictionary* defines "spontaneous ignition" as "the outbreak of fire in combustible material (as oily rags or damp hay), that occurs without application of direct flame or spark and is usually caused by slow oxidation processes (as atmospheric oxidation or bacterial fermentation) under conditions not permitting dissipation of heat."

This definition stresses an internal heat source or "self-heating" as a cause of ignition. But "spontaneous ignition" is also used to describe ignition at "low" temperatures caused by external heat sources other than direct flames or sparks. In practice, both mechanisms, or both heat sources (internal and external) may simultaneously contribute to ignition.

Minimum conditions for the spontaneous ignition of combustible solids such as wood are not simple. Generally, ignition requires adequate oxygen and adequately high temperature. Since oxygen is normally in abundance, usually only ignition *temperatures* are of concern. But there is no single critical temperature that is applicable under all circumstances.

A large amount of experimental work employing many different methods suggests that when small samples of wood are exposed to high temperatures for periods of minutes to hours, and in a few cases, days, ignition generally occurs for surrounding temperatures

as low as 400° to 500° F. This is generally considered the spontaneous ignition temperature of wood for "short" duration exposures. Since the minimum ignition temperature of wood smoke is substantially higher than this, it is evident that spontaneous ignition of wood starts with the combustion (or glowing) of charcoal, whose heat then may ignite the gases.

There is considerable field evidence that wood can ignite when exposed to much lower temperatures than 400–500° F. Many building fires have started with the ignition of wood in contact with hot water tanks or steampipes. In such systems pressure limiting devices should limit the temperature. Assuming that the maximum nominal pressure accurately indicates the maximum actual temperature, it appears that temperatures possibly as low as 212° F. and certainly as low as 250° F. have caused fires.

Unfortunately there is little laboratory evidence to corroborate such occurrences. Clearly the usual laboratory experiments, lasting no more than days, to measure spontaneous ignition temperatures are not necessarily applicable since years of exposure typically preceded the outbreak of fire in the field evidence. It is possible the field evidence is being misinterpreted, that the pressure limiting devices were not working properly, or the fires had other causes. But since most experts believe that temperatures at least as low as 250° F. can cause fires, I shall proceed on that assumption.

Possible contributing causes of ignition of wood at temperatures below 300° F. are self-heating, for which laboratory and field evidence exists, and special properties of wood charcoal, some not fully explored in the laboratory.

Internal or self-heating can be caused by biological processes, thermal degradation, or slow chemical oxidation. The maximum temperatures caused by these self-heating processes depend on the rate and density of heat evolution, and the rate at which heat can escape.

Biological processes require certain conditions. Low temperatures slow the life processes, and too high temperatures terminate them permanently. Sufficient moisture must also be present. In practice, oil-soaked rags and damp hay are especially prone to biological self-heating. Sound, dry wood such as is usually found in building construction is not very susceptible.

Thermal degradation (or pyrolysis) can be either an endothermic (heat-consuming) or exothermic (heat-generating) reaction. Thermal degradation is the breaking and forming of molecular bonds (or generally, chemical rearrangement) induced by heat and without the use or participation of oxygen from the air surrounding the substance. Conventionally pyrolytic reactions are thought to be endothermic at low temperatures and to become exothermic for temperatures above roughly 500° F., although it may be possible for the reactions to be exothermic at lower temperatures.

Slow chemical oxidation, as opposed to biological oxidation or fast chemical oxidation (combustion), probably occurs to some extent at almost any temperature. As with all chemical reactions, there is not a single minimum temperature at which the oxidation can proceed; the reaction is just slower at lower temperatures (even undetectable by casual observation). Rusting of iron is an example of a slow oxidation reaction. The heat evolved in slow oxidation reactions is often not noticeable because the rate is so low and the heat dissipates into the surroundings as quickly as it is generated.

Whatever the mechanism of self-heating, the effect becomes both more noticeable and potentially dangerous as the size of the sample increases. For example, a few straws of wet hay or one small oily rag will not spontaneously ignite, but in bulk, both can and often do. The reason is the insulating effect of the other portions of the pile. Although the heat-generating processes may be occurring throughout the sample, the center generally gets the hottest because it is the most insulated by the rest of the sample. If the temperature rise is sufficient, spontaneous ignition (or combustion) occurs (defined as either the glowing of charred material or the flaming of gases emanating from the sample).

Self-heating has occurred in wood fiberboard under laboratory conditions. When placed in a hot, constant, and uniform temperature environment, the samples became hotter than their environment, indicating self-heating. The highest temperatures were at the centers of the samples. Shown at the top of the next column are the results of these experiments with stacked wood fiberboard.

Thickness and approximate diameter of specimen (inches)	Lowest surrounding temperature for 5° self-heating rise (°F.)	Lowest surrounding temperature that resulted in ignition (°F.)
1	280	396
2	240	356
4	210	302
8	190	266
12	180	252
22	170	228

Surrounding temperatures as low as 147° F. caused perceptible self-heating of the 12-inch sample, but a 252° F. environment was required for the self-heating to be enough to cause ignition.

Fires clearly due to self-heating have started in stacked wood fiberboards, and piles of coal, coke, and charcoal. To avoid such fires, care must be taken to keep the size of piles or stacks down, to keep the ambient temperature down, to be sure that prepared or processed materials such as fiberboards, coke, and charcoal, are cooled before being stored in bulk, and to keep moisture away. After moistening, many materials seem to self-heat more than when kept dry.

Charcoal is of particular interest. Both laboratory and field evidence clearly indicate wood exposed to temperatures as low as 225° F. for months and years gradually carbonizes. Thus attention should probably be focused on the ignition of charcoal, not "natural" wood.

Unfortunately, charcoal is not simple. Its properties depend on its history. Fresh charcoal is capable of absorbing large quantities of gases, including oxygen. To what extent and under what circumstances the oxygen can oxidize some of the carbon is not well known, but it is conceivable that some self-heating could occur by this mechanism at low surrounding temperatures. Of course, self-heating generally can result in a significant temperature rise only if there is a sufficiently insulating environment around the reacting region. In the better known self-heating situations, this insulation is the material itself, in large thicknesses. Generally, the thicknesses of charcoal and wood are not large in building construction and only one side of the material may be exposed to the warm temperatures. However, building insulation on one side could contribute somewhat.

The key may be some special and as yet not fully explored properties of charcoal. For instance, moisture is important. Laboratory experiments have shown that the ignition temperature of moist charcoal is substantially lower than that of dry charcoal. In one experiment, moist charcoal ignited at 318° F., dry charcoal at 420° F.; 318° F. is getting down reasonably close to 200–250° F. Why moisture lowers ignition temperatures is apparently not well understood. The effects of time-varying conditions, both temperature and moisture content (or humidity of surrounding air) have not been explored in the laboratory. Conceivably charcoal may become especially reactive under some set of conditions, resulting in either a substantially lower than normal ignition temperature, or in sufficiently concentrated and rapid self-heating to achieve "normal" ignition temperatures.

There appears to be a gap between field experience and laboratory evidence concerning spontaneous ignition at low temperatures. A combination of more careful investigation of field evidence, and more laboratory work on ignition of wood charcoal is needed. In the meantime it seems prudent to accept the available field evidence at face value, that spontaneous ignition of normal thicknesses of wood can occur at temperatures at least as low as 250° F., and perhaps as low as 212° F.

However it is clear that ignition at these temperatures is rather rare; a fire is by no means guaranteed if wood reaches 250° F. In an experiment by the National Research Council of Canada, samples of wood and fiberboard were maintained at around 250° F. for 4 years. Considerable charring occurred, but ignition did not.

Another effect of long exposure of wood to elevated temperatures is loss of strength. Significant deterioration occurs at temperatures at least as low as 250° F. over a period of years. Many authorities feel temperatures as low as 200° F, could cause damage to wood as a structural material.

However arrived at, the values for maximum safe temperatures used by most safety organizations are

90° F. higher than ambient, or about 160° F.–170° F. for constant conditions for unexposed surfaces, such as beneath a floor protector;

117° F. higher than ambient, or about 185° F.–195° F. for constant conditions for exposed surfaces;

140° F. higher than ambient, or about 210° F.–220° F. for periods of time of up to about an hour.

175° F. higher than ambient, or about 245° F.–255° F. for occasional periods of time of up to about 10 minutes, such as during a short chimney fire.

These temperature limits are generally used for all common combustible materials except liquids, despite the fact that combustibility is a relative concept—all materials that can burn are not equally easy to ignite.

These limits are intended to contain a margin of safety. The actual numbers are somewhat arbitrary; they are not securely defensible by either field or laboratory data, but are basically educated guesses. See Bibliography for more details on experiments referred to in this Appendix.

APPENDIX 2

Clearances, Reduced Clearances, and Protectors For Residential Solid Fuel Heaters

The best evidence concerning the fire safety of clearances, and of the adequacy of wall and floor protectors, is field evidence—how many fires are caused in homes under normal "real life" conditions.

This information is difficult to obtain in a reliable and statistically significant form. Thus a second approach is usually taken. Well-entrenched estimates of how hot wood can get without danger of ignition are used. These estimates are based primarily on somewhat scanty field evidence as explained in Appendix 1. Using these estimates, it is possible to test equipment, clearances, and protectors in laboratories to see whether temperatures of nearby combustibles stay below dangerous levels. Both the temperature limits and the operating procedures for the appliances, such as how hot a fire is "reasonable," are in part arbitrary, but this approach is probably reasonable. We hope actual field accident studies and additional laboratory research will some day be available to permit a careful assessment of this method.

Not all equipment nor all possible installations will ever be tested in this manner. Thus the National Fire Protection Association and most building codes have installation prescriptions for unlisted equipment (equipment that has not been explicitly tested). In the interest of simplicity and easy enforcement, NFPA and codes give only a few specific clearances, protectors, etc., that are intended to cover all situations. They must be safe for the worst case. For instance, NFPA specifies 36 inches for the minimum clearance between a radiant stove and an unprotected combustible wall.

This is intended to be safe for all stoves, including of course the hottest and biggest. Therefore, this clearance distance is more than necessary for most other stoves, since smaller and inherently cooler stoves will not heat the wall as much.

A third method for assessing the safety of installations is theoretical—use of heat transfer theory to predict wall and floor temperatures near stoves. This is not a very useful method all by itself because there are too many unknowns, the most important of which is the temperature distribution on the outside of the radiating appliance.

However, theory is likely to give fairly accurate results for some *relative* effects on wall temperatures of different *size* stoves with the same surface temperature, or of different distances for the same stove. Some of the material in this Appendix has the following structure: NFPA's installation guidelines are taken as being safe, and theory is then used to predict other equally safe installations. In some cases theoretical considerations show inconsistencies in NFPA guidelines, indicating that the various parts of the guidelines do not achieve the same degree of safety.

This Appendix is intended as constructive and helpful criticism of NFPA and of building codes. As discussed in Appendix 4, writing codes and standards involves some very difficult compromises. It is hoped that this Appendix will increase general understanding of the principles involved in clearances and protectors, and thus increase the safety of solid-fuel heater installations.

Heat Transfer Theory

While the following discussion focuses on walls, it is also applicable to floors and ceilings, with minor modifications.

When a stove or other appliance, or stovepipe is installed near a wall, the wall is usually hottest directly behind the stove, and the hottest part of the wall is its surface facing the stove. The wall receives radiation from the stove, and from the rest of the room. The wall surface loses heat by radiation into the room, by convection to the room air, and by conduction back into and through the wall (Figure A2–1). I assume that there is no lateral (sideways) conduction within the wall, that there are no energy gains or losses due to chemical reactions (pyrolysis of wall) or phase changes (drying of wall), that the emittance of the appliance is unity, and that the emittance (and absorbance) of the wall has no wavelength dependence in the spectral region of interest, and that its directional dependence follows the cosine law of ideal black surfaces. The room's emittance is effectively unity since it fully encloses the hot wall surface.

The steady-state wall surface temperature will be such that the heat loss rate from the wall's exposed surface balances its heat gain rate from the stove and room. Per unit surface area, we have

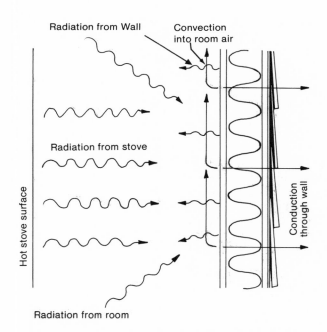

Figure A2-1. Heat gains and losses at a wall surface exposed to radiation from a stove or other radiating surface. Since the losses from the wall surface all increase with surface temperature, the balancing of the gains and losses determines the equilibrium surface temperature. Lateral conduction within the wall is not illustrated but can contribute by taking heat away from the hottest region directly behind the stove.

Gains:

Radiant, from stove $F\sigma\varepsilon_w T_s^4$

Radiant, from room $(1 - F)\,\sigma\varepsilon_w T_R^4$

Losses:

Radiant, to room $\sigma\varepsilon_w T_w^4$

Convective, to room air $h'\,(T_w - T_{RA})^{4/3}$

Conductive, back into wall $U_w\,(T_w - T_{w0})$

where

F = configuration factor of the radiating appliance and/or stovepipe as seen from a point on the wall

σ = Stefan-Boltzmann constant, 1.7×10^{-9} Btu/hr ft^2 (deg R)^4 (5.67×10^{-8} watts/m^2 (deg K)^4)

T_s = the appropriate mean stove surface temperature (with hot regions receiving higher weighting, in keeping with the T^4 dependence of radiation per unit surface area)

T_R = room surface temperature

ε_w = emittance of the (hot) wall, estimated value 0.9

T_w = temperature of the (hot) wall

h' = natural convection heat transfer coefficient, excluding temperature dependence, estimated value 0.19 Btu/hr/ft^2 (°F.)

T_{RA} = temperature of room air in vicinity of (hot) wall

U_w = U-factor for wall (heat flow per unit area per unit temperature difference)

T_{w0} = surface temperature on other side of wall.

The balancing of gains and losses in steady state implies

$$F\sigma\varepsilon_w T_s^4 + (1 - F)\sigma T_R^4 = \sigma\varepsilon_w T_w^4 + h'\,(T_w - T_{RA})^{4/3} + U_w\,(T_w - T_{w0})$$

For a square stove side with edge length e a distance d from a parallel wall,

$$F = \frac{4}{\pi}\,\frac{1}{\sqrt{1 + (2d/e)^2}}\,\arctan\frac{1}{\sqrt{1 + (2d/e)^2}}$$

for the point on the wall directly opposite the center of the stove's side (Figure A2-2). For stove sides with other reasonably normal shapes, F is apparently the same as for a square side with the same area.

For straight runs of stovepipe, as seen from the point on the wall opposite the midpoint of the pipe,[2]

1. See, for instance, H. C. Hottel, "Radiant Heat Transmission," *Mechanical Engineering* Vol. 52 No. 7 pp 699–704 (1930).
2. D. I. Lawson, L. L. Fox and C. T. Webster, "The Heating of Panels By Flue Pipes," Fire Research Special Report No. 1, Department of Scientific and Industrial Research, Her Majesty's Stationery Office (1952).

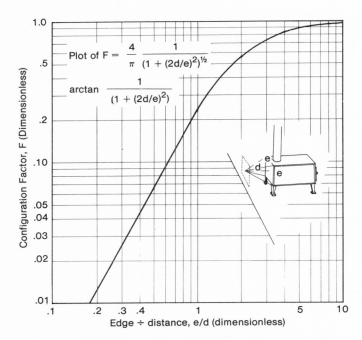

Figure A2-2. The configuration factor of a square surface with edge length e at a distance d from a parallel wall, as seen from the point on the wall directly opposite the center of the square surface. Rectangular surfaces with edge ratios between 0.5 and 2 have approximately the same configuration factor as a square surface with the same area.

$$F = \frac{2}{\pi} \left[\frac{ab}{(a^2b^2 + (b^2-1)^2)^{\frac{1}{2}}} \arctan \frac{(b^2-1)^{\frac{1}{2}}}{(a^2b^2+(b^2-1)^2)^{\frac{1}{2}}} + \frac{1}{b} \arctan \frac{a}{(b^2+1)^{\frac{1}{2}}} \right]$$

where

a = (length of pipe) ÷ (diameter of pipe)
b = (distance from wall to *center* of pipe) ÷ (radius of pipe).

This function is graphed in Figure A2-3.

Effects of Appliance Size and Surface Temperature on Minimum Safe Clearances

Figures A2-4 through A2-7 give results of this theory for $\varepsilon_w = 0.9$, $T_R = T_{RA} = 70°$ F., $U = 0.1$ Btu/hr/ft^2 °F. and $T_{w0} = 30°$ F. The predicted wall temperatures are probably overestimates due to the intentional conservatism of the assumptions. The available evidence seems to confirm this (see Bibliography). However the primary objective here is to illustrate relative effects. (Note that in comparing these predictions with experimentally measured values, stove surface temperatures must be proper averages as defined previously, and true *surface* temperatures must be obtained for both the radiating and absorbing surfaces. Finally,

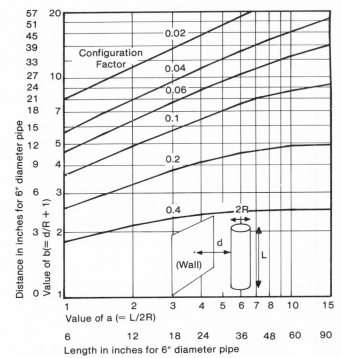

Figure A2-3. The configuration factor for stovepipe with length L and radius R at a distance d from a parallel wall, as seen from the point on the wall directly opposite the midpoint of the pipe.

wall conductances vary considerably, and when they are large, T_{w0} is not closely approximated by the air temperature on the back side of the wall.)

Using these figures, the following conclusions may be drawn. A clearance of 36 inches from a combustible wall may be inadequate for the hottest and largest stoves. A 28 × 29-inch stove side, such as on the Riteway Model 2000, has a configuration factor of 0.165 as seen from 36 inches away directly opposite the center of the side. Taking 200° F. to be the maximum safe wall temperature, the theory predicts that this temperature will be reached for an average stove surface temperature of about 600° F. without stovepipe radiation, or about 570° F. with a hypothetical top-connecting stovepipe.

Thus this theory suggests that the standard 36-inch clearance may be inadequate for stoves significantly larger than the Riteway Model 2000 if their average surface temperature can be as high as about 600° F. and for stoves about the size of the Riteway if their average surface temperature can significantly exceed about 600° F. (Because of the conservatism of the theory, in practice these temperatures might be roughly 100 to 200 degrees higher.)

I suspect such circumstances are rare. Temperatures as high as 600° to 800° F. are probably very rare as an *average* over the whole side of a large stove.

Even if such high temperatures can be realized, they would probably not be sustained continuously in ractice. Thus the wall would not necessarily achieve the predicted 200° F. steady-state temperature. In addition, the 200° F. limit itself may be conservative and is generally intended for steady-state exposure, with higher temperatures being allowed for short times.

At the other extreme, for a small stove such as the Jøtul 602, 36 inches of clearance is much more than necessary. The back of the Jøtul 602 is about 12×12 inches, with a configuration factor of 0.034 at 36 inches. From Figure A2–4 it can be seen that the stove's surface temperature would have to be about 1100° F. for wall temperatures to exceed 200° F. with the 36-inch clearance. If stovepipe radiation is included, stove and pipe average surface temperatures (assumed to be the same) would be about 800° F. Again, these estimates are intended to be conservative. Such high average surface temperatures are almost impossible to achieve in any normal use of the stove. Rough estimates of safe clearances from combustible walls for stoves, stovepipes, and stoves with vertically connecting stovepipes, are given in Figures A2–5 through A2–7. The principal assumption is that 200° F. is the maximum safe steady-state temperature for the wall surface. It is clear from this figure how very

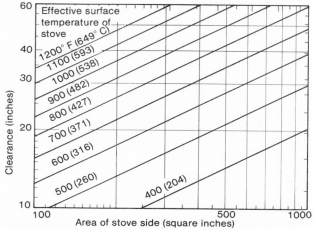

Figure A2-5. Theoretical safe clearances of radiant stoves by themselves (without stovepipe contributions) from unprotected combustible walls, where "safe" means wall temperatures will not exceed about 200° F, at an ambient room temperature of 70° F.

important both stove surface area and stove surface temperature are to minimum safe clearances.

The dashed line in Figure A2–7 is constructed to have similar slope to the other curves and to give an NFPA-recommended clearance of 36 inches for the side of the Riteway Model 2000, a relatively large radiant stove. Presuming this stove is safe at 36 inches and that its maximum credible average surface temperature is as high as that for any stove, the dashed curve gives equally safe clearances for smaller stoves. Clearly any single clearance from combustibles for all stoves is bound to be irrational for many stoves. A Jøtul 602 with its back towards an unprotected combustible wall is as safe at a clearance of about 20 inches as is a Riteway 2000 at 36 inches with its side facing the wall, assuming equal surface temperatures for the stoves.

Applied to stovepipe alone, these theoretical considerations give support to the three-times-diameter rule over the 18-inch rule for stovepipe clearance. For long stovepipe runs, all sizes of stovepipe with the same surface temperature result in the same wall surface temperature when the clearances are three times the pipe diameter. However, with a uniform clearance of 18 inches this is not the case. For an equal surface temperature, chosen to give a wall temperature of 200° F. with a 6-inch diameter pipe, a 4-inch pipe at the same 18-inch distance would make the wall about 169° F. and a 10-inch pipe would make the wall 247° F.

These considerations indicate some areas where codes might be made more reasonable but alone are not adequate justification. Empirically, there may be a correlation between stove (or stovepipe) size and surface temperature. Both field and laboratory data are needed to pin down these theoretical estimates. Even if substantiated by experiments, these relations should not automatically be incorporated in codes. As discussed in Appendix 4, codes must be compromises between rationality and simplicity.

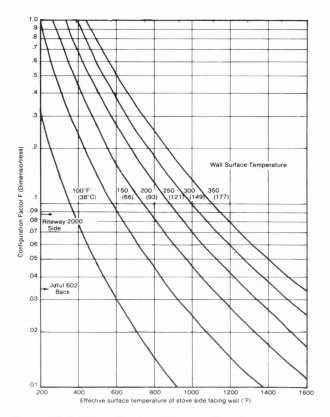

Figure A2-4. Theoretical maximum wall surface temperature directly opposite a hot appliance or stovepipe as a function of the effective black body radiant surface temperature and the configuration factor. The assumptions are intentionally conservative with respect to safety implications.

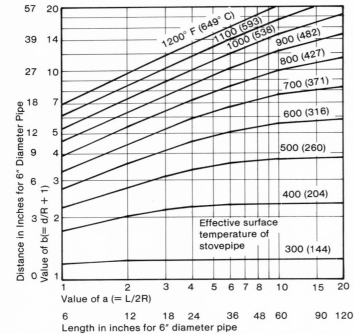

Figure A2-6. Theoretical safe clearances of stovepipe from unprotected combustible walls, where "safe" means calculated wall temperatures will not exceed about 200° F. The predicted safe clearances are intended to be conservative.

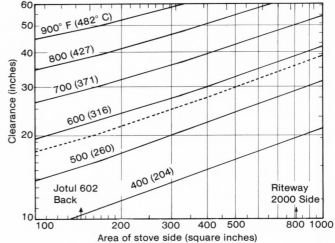

Figure A2-7. Theoretical safe clearances of stoves with vertically connecting stovepipe, where "safe" means the maximum calculated wall temperatures will not exceed about 200° F, at an ambient room temperature of 70° F. The pipe is assumed to be very tall, to have a diameter of 8 inches, to have an emittance of 1 and to be at a uniform temperature equal to that of the stove surface facing the wall. The location of the hottest (200° F.) point varies from being opposite the center of the stove to being opposite the top of the stove. The dashed line is constructed to have the same slope as the other lines and to give the NFPA-recommended clearance of 36 inches for one of the larger radiant stoves, the Riteway model 2000. (These clearances are intentionally conservative.)

Desirable Thermal and Radiative Properties for Wall Protectors

Wall protectors for reduced clearances should be designed to keep the wall cool. The most effective of common types is essentially a ventilated air space. The thermal properties of the panel or construction used to create the air space are not usually critical, but the following thermal properties are desirable (these also apply to non air-gapped protectors):

1. High infrared reflectance on the front of the panel. Although it may not be wise to *rely* on reflectance of an exposed surface, since it may tarnish or be painted and hence lose this property, it can be so effective that the air gap is not needed. Polished bare copper (it must not even be covered with a clear paint) can be *cool* to the touch when only a foot from a hot stove or stovepipe.

2. Low thermal conductance through (perpendicular to) the panel.

3. Low infrared emittance on the back side of the panel to minimize radiation to the wall. High reflectance and low emittance (at the same wavelengths) always occur together; thus polished metals, such as copper, are also poor emitters. (Low emittance is not as important on the back side of protectors without air gaps.)

4. High lateral (sideways) conductance within the panel. This helps by carrying the heat away from the hot regions of the panel to the cooler perimeter, thus decreasing the maximum temperatures directly behind the stove. Lateral conductance serves to average out panel temperatures, which lowers the maximum temperature.

Since few practical materials have anisotropic thermal conductivity, choosing a panel material with high lateral conductance generally implies high conductance through the panel as well. The best balance between low perpendicular conductance and high lateral conductance has not yet been determined, but it is possible that high-conductivity materials could be superior in some situations, especially those with ventilated air gaps.

One practical way to increase lateral conductance and simultaneously decrease perpendicular conductance is by increasing the thickness of a thermally isotropic protector material. Thus it is doubly useful to increase the thickness of any given protector material.

5. High mass in a panel contributes to limiting peak temperatures of the wall during temperature surges of the appliance or stovepipe due to the panel's heat storage capability.

The Inadequacy of Insulation and Other Non-Air-Gapped Protectors

Most building codes and NFPA (in "Heat Producing Appliance Clearances 1976," NFPA 89M-1976) allow substantial reduction in clearances with the non–air–gapped protector consisting of 0.027-inch sheet metal on a 1-inch mineral wool batt reinforced with wire or equivalent. When the unprotected required clearance is 18 inches, use of this protector permits clearances of 3 to 4 inches; an unprotected clearance of 36 inches may be reduced to 12 inches.

I would advise against such large reductions in clearances with this protector, in particular when the protected surface is an insulated wall or ceiling. A typical insulated exterior wall in residential construction has a thermal resistance of roughly 12 $(Btu/hr\ ft^2\ ^\circ F.)^{-1}$. A typical 1-inch thick mineral wool blanket has a thermal resistance of roughly 3. The sheet metal has negligible resistance, and if painted, as NFPA allows, has negligible infrared reflectance. Thus when the protector is placed on the wall, the total resistance is roughly 15. Since the protector constitutes only about 20 percent of the resistance, only about 20 percent of the temperature difference across the whole structure will occur across the protector. Thus the protector does little to keep a well-insulated wall cool.

Continuing this example, NFPA guidelines allow a radiant stove to be placed 12 inches from a wall with this type of protector. If the stove were a Riteway Model 2000 with its side facing the wall, the predicted maximum temperature of the surface of the protector would be about 400° F. with a stove surface temperature of 700° F. With an outdoor wall surface temperature of 30° F., the total temperature drop would be 370° F. With 20 percent of this drop occurring across the protector, the original wall surface would be at about 325° F., well above the accepted maximum safe temperature of about 200°.

Of course for an uninsulated wall, the insulated protector can be quite significant. For a wall consisting of only ¾-inch wood, the thermal resistance is about 1. Thus the total resistance of the protected wall would be about 4. Consequently about three-fourths of the total temperature difference across the whole structure would occur across the insulation. With the same assumptions as in the previous example, the wood wall surface temperature would be only about 125° F.

I suspect the reason NFPA allows such small clearances for this type of protector is due to an oversight many decades ago. The original research on which most of NFPA's clearances seem to be based is reported by J. A. Neale in "Clearances and Insulation of Heating Appliances," Bulletin of Research No. 27 (Underwriters Laboratories, Inc., 1943). Bare ½-inch pine boards were used to represent walls and ceilings. No insulation was applied to their back sides. Thus protectors consisting of insulation applied to the

exposed sides of the boards were found to be very effective. What seems to have been neglected somewhere along the way is that many walls in buildings are insulated and thus these experimental results are not directly applicable.

There may be similar but less dramatic oversights in NFPA's reduced clearances with non–air–gapped sheet metal on ¼-inch asbestos millboard protectors, both for walls and for floors. The problem is not quite so serious because NFPA-allowed reduction in clearances is not as substantial. In addition, particularly when used as a wall protector, this construction may be more likely to be installed with a slight unintentional ventilated air gap. But the same general criticism seems valid—that the original tests, and NFPA's recommendations, seem to be based on uninsulated walls and floors that require much less protection than would insulated walls and floors for the same degree of protection.

Test results on ventilated-type protectors are not as much affected by whether or not the tested wall is insulated. Heat removal by convection is so dominant that the wall's thermal resistance is not of as great importance.

It is still common laboratory practice, as required by the product standards (e.g., UL 103, 127, 737 and 1482), to use uninsulated walls, ceilings, and floors around stoves, fireplaces, and chimneys when testing these systems for safe temperatures and clearances. The clearances thus determined are less than would be the case if the test structure were insulated as are many walls and ceilings in homes. It is my hope that this discrepancy is not dangerous. The margins of safety already incorporated in these tests may be more than enough to accommodate this discrepancy.

In summary, it seems clear that some testing is needed on the effect of wall, ceiling, and floor construction on the effectiveness of various protectors. In particular highly insulated constructions should be tested since these probably represent the worst cases, requiring the most protection for reduced clearances. The results may suggest a need to modify our building codes and product safety standards.

In the meantime, I recommend using only ventilated-type wall and ceiling protectors, and either removing any insulation in the floor under a stove or fireplace or using a ventilated floor protector.

Extents of Wall and Ceiling Protectors For Stovepipe

For installation of stovepipe chimney connectors with reduced clearances to combustible walls or ceilings, NFPA specifies that the panel used to protect the wall or ceiling be big enough so that all of the wall or ceiling surface that is closer than the allowed unprotected clearance is covered by the protector. Thus, using NFPA's standard 18 inches for the required

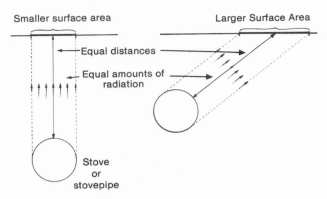

Figure A2–8. The geometrical basis for the effect on wall temperature of the angle of incidence of the radiation. A surface will get hottest when facing directly at the source of radiation. When a surface is tipped somewhat away from the radiation source, the same quantity of radiation is absorbed over a larger surface area; thus the surface does not get as hot.

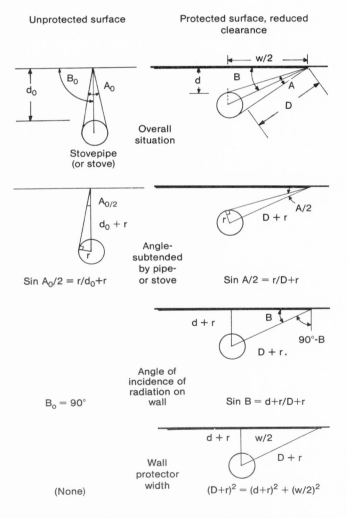

Figure A2-9. Geometrical considerations for calculating radiant flux on a wall from stovepipe or a spherical or cylindrical stove.

clearance to unprotected walls, in reduced clearance installations the protector must cover all wall area within 18 inches of the stovepipe. (I am assuming here straight stovepipe and a planar and parallel wall surface.)

It is my opinion that this specification results in unnecessarily large protective panels. The reason is that no account is taken of the angle of incidence of the radiation on the wall. When the 18 inches is measured out to the wall beyond a protective panel, the grazing angle of the heat radiation from the pipe results in less radiation hitting the wall *per unit area of wall surface*, and hence the wall does not get as hot as it would directly behind a pipe at 18 inches (Figure A2–8). Beyond a protector panel, the wall does not directly face the stovepipe.

The radiant intensity on the wall from a stovepipe is proportional to the product of the angular area of the pipe as seen from the wall, and the cosine of the angle of incidence of the radiation striking the wall. Using the notation in Figure A2–9, we have

Radiant intensity on wall $\propto$ (angular area of pipe) $\times$ (cosine of angle of incidence)

$$\propto A \times \cos(90 - B) \qquad \propto \frac{A}{2} \times \sin B$$

Here I have assumed that the effective angle of incidence is the average over the actual distribution of angles. This slightly overestimates the flux on the wall directly behind a pipe and this leads to conservatively generous calculated wall protector widths. In assuming the angular area of the stove to be proportional to the angle A, I am also being conservative—I am essentially assuming the stovepipe is *very* long.

Assuming that the needed extent of a wall protector is such that the exposure of the wall (radiant intensity per unit wall area) just beyond the protector is equal to the allowed exposure directly behind the pipe when it is at the minimum allowed distance from an unprotected wall, we obtain

$$\frac{A_0}{2} \sin B_0 = \frac{A}{2} \sin B$$

Making substitutions from Figure A2–8, we find

$$\arcsin\left(\frac{r}{d+r}\right) = \frac{d+r}{D+r} \arcsin\left(\frac{r}{D+r}\right).$$

Solving this equation for D and then computing the required wall protector width $w = 2\sqrt{((D+R)^2+(d+r)^2)}$ yields the results shown in Table A2–1. Considerable reduction in NFPA wall protector extents is possible without any loss in safety. For example, application of NFPA's recommendations would result in a 34.5-inch wide wall protector for 6-inch pipe at a reduced clearance of 9 inches. My calculations indicate a 20.9-inch width would be adequate. Instead of 18 inches out

140

Table A2-1. Protector Widths for Stovepipe

Widths of Wall and Ceiling Protectors for Reduced Stovepipe Clearances, According to NFPA and to the Author's Calculations. The Author's Recommendations Take into Account the Angle of Incidence of the Radiation from the Stovepipe on the Wall or Ceiling.

PIPE DIAMETER (INCHES)				8					6					4		
Minimum Unprotected Clearance According to NFPA (Inches)				18					18					18		
Minimum Clearance With Protection (Inches)		Min. Distance to Unprotected Surface		Min. Protector Width[5]		Min. Distance to Unprotected Surface		Min. Protector Width[5]			Min. Distance to Unprotected Surface		Min. Protector Width[5]		
	NFPA (1)	A (2)	NFPA (3)	A (4)	NFPA (1)	A (2)	NFPA (3)	A (4)		NFPA (1)	A (2)	NFPA (3)	A (4)		
12	18	14.8	30.2	19.7	18	14.8	29.4	19.0		18	14.8	28.6	18.6		
9	18	12.9	35.5	21.6	18	12.9	34.5	20.9		18	12.8	33.4	19.8		
6	18	11.6	39.2	22.1	18	10.8	37.9	20.9		18	10.7	36.7	19.7		
3	18	8.5	41.7	20.7	18	8.3	40.2	19.2		18	8.0	38.7	17.3		

1. Minimum distance to unprotected wall according to NFPA.

2. Minimum distance to unprotected wall according to the author's calculations.

3. Minimum wall protector width, according to NFPA (computed from column 1).

4. Minimum wall protector width, according to author's calculations (computed from column 2).

5. These numbers in 3 and 4 represent the *wall* width needing protection. The protector itself can typically be a few inches narrower if the protector protrudes or is spaced out from the wall because the protector shadows more of the wall (from the pipe's radiation) than it covers.

to unprotected wall at the edges of the protector, 12.9 inches is adequate, offering safety equal to that of a 6-inch pipe 18 inches from an unprotected wall.

The numbers in Table A2–1 indicate that very nearly the same protector width is required for all cases considered. Thus there is a possibility here of NFPA's recommendations being modified to become both more rational *and* simpler simultaneously. In addition, wall protectors would be less expensive. The additional simplicity would come from specifying the protector panel width directly. A conservative width for all stovepipe protectors for pipe as large as 8 inches in diameter would be 24 inches, although 20 inches might also prove to be adequate. It is conceivable that the wider extent is needed to help dissipate the heat absorbed in the center of the panel, thus keeping its temperature lower. However, this seems unlikely since the NFPA-recommended protective panels have such low conductances. Wider panels might also be necessary near the stove where the total radiation hitting the wall comes from both the stovepipe and the stove. Experimental verification of all these results is, of course, necessary.

Extents of Wall Protectors for Appliances

Wall protector extents for *stoves* and other solid fuel appliances are unnecessarily large as specified by NFPA and as required by most building codes. The reason again is the apparent neglect of the smaller angle of incidence of the heat radiation on the wall at locations other than directly behind the appliance. If 36 inches is a safe clearance between a stove and an unprotected combustible wall, to require a clearance-reducing wall protector to extend so far out and up that no wall surface within 36 inches is unprotected is excessive. Out at the edge of the protector, the grazing angle of incidence means each square inch of wall area is hit by a smaller amount of radiation than if it were oriented to face directly towards the appliance.

A general theoretical treatment of this effect of angle of incidence is more difficult than in the case of stovepipe chimney connectors because of the variety and complexity of appliance shapes. One can obtain a rough idea of the effect by assuming the appliance to be spherical in shape, not because many stoves are spherical, but because the computations are easier for a spherical shape, and the overall nature of the results is not critically dependent on shape within certain reasonable limits.

The calculations are similar to those for stovepipe except that the angular area subtended by the appliance is proportional to the *square* of the angle it subtends. Using the same notation but where r is here the radius of the spherical appliance rather than the radius of a cylindrical stovepipe, we have

$$\text{radiant intensity on wall} \propto \left(\frac{A}{2}\right)^2 \times \sin B.$$

The condition of equal intensities directly behind an appliance at the no-protection clearance, and just beyond the protection in the case of reduced clearance becomes

$$\left(\arcsin \frac{r}{d+r}\right)^2 = \frac{d+R}{D+R}\left(\arcsin \frac{r}{D+r}\right)^2.$$

Solutions for D and the resulting wall protector dimensions are given in Table A2–2. It can be seen that NFPA's guidelines result in protector panels being about 2 feet wider (and 1 foot taller) than necessary, where "than necessary" means for wall temperatures equivalent to what NFPA allows for a stove 36 inches away from a wall without a protector.

There is also limited experimental evidence that these reduced protector extents are safe. Testing done by an independent laboratory for Fisher Stoves International, Inc. has shown that for many of the Fisher stoves (not fireplace stoves) the wall protector extension of only 18 inches beyond the projection of the stove on the wall is more than enough in order for the stoves to pass the Underwriters Laboratories safety standard (UL 1482) for stoves installed with a reduced clearance of 12 inches.

For fireplace stoves the wall protector extents in some cases should be larger. If the stove is installed so that direct radiation from the fire can hit the protected wall, the independent laboratory found that a 30-inch protector extension beyond the stove was satisfactory.

Summarizing these theoretical and limited experimental results, I suggest the following rule of thumb:

Wall protectors for stoves and fireplace stoves should extend 18 inches beyond the (perpendicular) projection of the stove on the wall, except in installations where the protected wall could receive radiation directly from the fire of an open fireplace stove, in which case the protector should extend 30 inches beyond the projection of the fireplace stove into the wall region receiving direct radiation. (See Figure 2–48.)

An advantage of this prescription compared to NFPA's is that it specifies widths directly. It is thus easier to use in practice. It also permits use of smaller panels. Again, experimental confirmation is important; it may be that protection against both stove and stovepipe radiation together requires larger protectors.

Extents of Floor Protectors

There is an apparent discrepancy between floor and wall protector extents as recommended by NFPA. According to NFPA, floor protectors need only extend

Table A2-2. Wall Protector Extents for Appliances[1]

Appliance Diameter (2r) (in)	Reduced Clearance Between Appliance and Wall (d) (in)	Minimum Distance to Exposed Wall (D)		Minimum Width of Wall Protector (W)		Minimum Wall Protector Extent Beyond Projection of Appliance ½(W−2r)	
		NFPA (in)	Author (in)	NFPA (in)	Author (in)	NFPA (in)	Author (in)
12	12	36	25.7	75.9	52.2	32.0	20.1
24	12	36	26.3	83.1	57.9	29.6	17.0
36	12	36	26.7	89.8	66.3	26.9	15.2
48	12	36	27.0	96.0	72.2	24.0	12.1

1. These numbers represent the *wall* width needing protection. The protector itself can typically be a few inches smaller because (and to the extent that) it protrudes out from wall since it shadows (with respect to the stove's radiation) more of the wall than it covers.

These extents take into account the angle of incidence of the infrared radiation from the appliance on the wall. The results are theoretical, not experimental. The safe clearance with no protection (d_0) is assumed to be 36 inches, which is more than necessary for small stoves.

6 inches beyond the sides of a stove, on sides without doors. Wall protectors must extend at least 2 feet beyond the projection of the stove onto the wall.

This large difference in protector extents is inconsistent in light of the fact that the radiant heat flux from the stove side onto the nearby floor and onto the nearby wall can be roughly the same. For a cubical stove with uniform surface temperature, with 12-inch legs, and placed at the reduced clearance of 12 inches from a combustible wall, the radiant flux on the floor and on the wall (or its protector) will be about the same at equal distances from the center of the side of the stove (Figure A2–8). Radiation travels equally well in all directions, including down.

There are a number of possible interpretations of the difference in extents of wall and floor protectors recommended by NFPA. It may be that floor protector extents are inadequate against radiant heating from the stove's side and bottom. I have observed slightly darkened areas on wood floors to the sides of a stove. These may have been due to radiant overheating. I have also heard reports of wood floor finishes catching fire apparently due to downward and outward radiation from stoves.

Another possible interpretation is that NFPA's wall protector extents are unnecessarily large. There is

theoretical and experimental evidence for this, as discussed previously.

I do not contend the extents should necessarily be the same even when the geometry is symmetric (leg length equal to clearance from wall, and a cubical stove). Walls usually need more protection for the following reasons: stove sides and backs are usually hotter than stove bottoms; walls are often exposed to more radiation from chimney connectors than are floors; and air temperatures near the wall are likely to be higher than near the floor. However, the difference between 6 inches and 2 or more feet for extents beyond floor and wall projections of the stove seems too large, particularly since the 6-inch extent of floor protectors applies no matter what the leg length is. If the stove has short legs, the stove side is closer to the floor than to the wall.

Again, there is supportive experimental evidence from independent testing laboratories. Many stoves would not pass the floor temperature limitations in the proposed standard UL 1482 with only 6-inch extensions of floor protectors beyond their sides. However, there is not yet enough evidence to be able to specify a general recommendation for all stoves.

I expect a floor protector extension of 12 inches beyond doorless sides is adequate for most stoves, but there may be some exceptionally big and/or hot and/or short-legged stoves requiring more space. I suspect that no stove on the market today requires more than 18 inches.

Floor Protector Materials

Asbestos millboard is a material commonly listed in codes and standards for protection of combustibles near heat-producing appliances. However, because of the possible health hazard due to asbestos fibers, and due to the lack of easy availability of the material, alternatives are needed.

Actual testing of floor and wall protectors is the most reliable way to determine their adequacy. Since little such testing has been published and since not all systems will ever be tested, a theoretical treatment can be useful to serve as an approximate guide.

NFPA recommends ¼-inch asbestos millboard covered with 0.013-inch sheet metal as a floor protector under stoves with legs between 6 and 18 inches long (NFPA HS-10). The same combination of materials is one of NFPA's recommended wall and ceiling protectors for reduced clearances.

Equivalent protectors have equal thermal resistance. The thermal resistance of sheet metal on asbestos millboard is not uniquely determined by the NFPA description; emittances/absorbances affect hear transfer, as does the thickness of the two small air gaps. These gaps will be affected by the flatness of the floor, the millboard, and the sheet metal, and by the weight distribution of the stove. In computing equivalent protection systems, it is reasonable to assume the worst

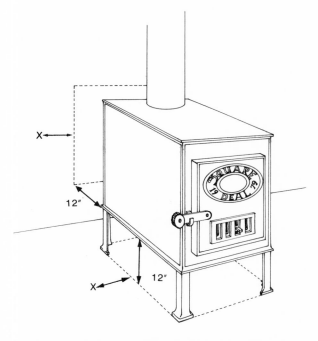

Figure A2–10. An example illustrating the rationale for more equal floor and wall protector extents than are now recommended by NFPA. The two locations marked "X" receive approximately equal amounts of radiation from the stove's side. Yet NFPA's recommended floor and wall protector extents are 6 inches and about 2 feet respectively, measured from the stove's projections (dashed lines in the figure).

case that meets NFPA's description—namely negligible thermal resistance due to the two air gaps and an absorbance of 1 for the exposed surface.

The conductivity of asbestos millboard is about 0.07 Btu/hr ft^2 °F.[1] at 85° F., and may rise to on the order of 0.1 at 200° to 300° F. Using the value 0.1, the resistance of ¼-inch asbestos millboard is about 0.2 hr °F. ft^2/Btu (R = thickness/conductivity = 0.02/0.1). The sheet metal's resistance is negligible. Two air gaps each 0.010 inch thick and with surface emittances of unity would increase the resistance by roughly 50 percent.

Alternative materials often used as floor protectors are listed in Table A2–3. Making the same worst-case assumptions of negligible air gaps and no surface reflectance, one can see from the table that solid masonry, with roughly one-fourth the thermal resistance of asbestos millboard, needs to be about four times as thick, or 1 inch thick, to give the same protection as ¼-inch asbestos millboard. Roughly ½-inch of sand is also equivalent. One-fourth inch of asbestos cement board is clearly not equivalent.

For practical, conservative and simple recommendations, I would suggest 4 inches of solid masonry or 2 inches of sand or small gravel. This should be at least equivalent to ¼-inch asbestos millboard plus sheet metal even if some credit is given to the air gaps. It should also be at least equivalent to ⅜-inch asbestos millboard plus sheet metal, another protector often used as a standard.

These equivalencies are based on *steady state* conductances *through* the protectors. Thick masonry has two additional advantages. Its mass offers additional protection during time-varying conditions, and

---

1. *Marks Standard Handbook for Mechanical Engineers*, 7th edition, (McGraw-Hill, 1967), p. 4–95.

Table A2-3. Thermal Resistances

Thermal Resistances Per Inch Thickness for Some Materials Used for Floor Protection Under Stoves

Asbestos millboard	0.8–1.2 hr°Fft^2/Btu
Asbestos cement board	0.2–0.3
Sand or gravel	0.4–0.8
Brick	0.1–0.3
Limestone	0.1–0.2
Concrete	0.13–0.17
Sandstone	0.07–0.12
Marble	0.05–0.07
Granite	0.04–0.08

its higher transverse conductance, if the masonry is monolithic or the masonry units are mortared, is more effective at spreading heat out from potential hot spots, thus diminishing peak temperatures.

For wall protection, 4 inches of masonry, such as a brick facing, without an air gap should allow radiant stoves to be placed 24 inches from a combustible wall, since NFPA allows this reduced clearance with a non-spaced-out sheet metal on ¼-inch asbestos millboard protector.

These alternative floor and wall protectors offer at least as much protection as the NFPA-recommended sheet metal on ¼-inch asbestos millboard. As discussed previously, this does not necessarily mean they are safe. Because NFPA's recommendations seem to be based on tests done on uninsulated floors, I would recommend either removing any insulation from under the floor under the wood heater when NFPA's or any equivalent protectors are used. If insulation is left in, I would recommend using a ventilated floor protector. It may also be wise to remove insulation in floors under factory-built fireplaces.

APPENDIX 3

The Safety of Chimneys

Masonry Chimneys

Ever since prefabricated chimneys built of metal and other materials were introduced, there has been controversy about the relative merits of these "new" chimneys compared to the traditional lined masonry chimney. Because the prefabricated chimneys were a radical departure from tradition, they were scrutinized closely by safety authorities. It was decided these chimneys should have to pass a safety test. A test procedure (UL103) was developed by Underwriters Laboratories, Inc., and essentially the same test is used today.

An important part of the test consists of sending hot flue gases through an installed chimney and measuring temperatures on nearby wood, looking for dangerous conditions. Excessive temperatures would indicate the need for more clearance in the installation, or a change in chimney design. The contemporary version of this test involves (approximately): using 1000° F. flue gases for as long as it takes for all temperatures of the chimney and the surrounding building structure to reach their maximum values; using 1400° F. flue gases for one hour, and using 1700° F. flue gases for ten minutes. These test conditions are intended to represent the worst case conditions that would occur in *most* actual field installations most of the time. This is the appropriate philosophy for a safety test. The actual numbers are somewhat arbitrary and are basically educated guesses as to the most severe conditions normally encountered, with the statistical meaning of the word "normally" not defined.

During these tests, any wood near the chimney must not get too hot if the chimney is to pass the test. The temperature limits are different for each part of the test. Temperatures of wood may not increase over ambient (room) temperatures by more than 117° F., 140° F., and 175° F., respectively. If ambient temperature is 70° F., the maximum allowed temperatures are 187° F., 210° F., and 245° F. during each segment of the test. All listed prefabricated chimneys on the market today pass these tests.

At the time when this safety performance standard was being developed, people naturally wondered whether masonry chimneys would pass the same test. Experiments in a number of laboratories indicate quite conclusively the answer is "No." (See references in Bibliography.) With 1000° F. flue gases, measured in the stovepipe connector five chimney diameters from the bottom of the chimney, typical temperature rises on joists spaced 2 inches from the masonry chimney are found to be roughly 200° to 300° F.—on the order of two to three times the 117° F. rise allowed for prefabricated metal chimneys!

Three possible conclusions are that masonry chimneys are unsafe, that the standard is too strict, and that the standard should not apply to masonry chimneys.

1. *Are masonry chimneys unsafe?* It is almost sacrilegious to ask the question. Masonry chimneys are the standard, or at least have been for a long time. But they came to be the standard by default—they were the only competitors on the field. This, of course, does not imply they are as safe as they should be.

Safety is of course a matter of degree. There is no such thing as a perfectly safe system. The best way to determine the safety of a product is through investigation and statistical analysis of actual field accidents. I am not aware of such a study on chimneys. It is needed, for there are reports of fires started "by" masonry chimneys built to contemporary code standards. All types of chimneys could be made safer, but whether up-to-code masonry chimneys are now a serious weak link in the safety chain of solid-fuel heating is not yet known to the best of my knowledge.

Chimney standards in many other Western countries are more conservative than in the United States. The Canadians do not permit *any* combustible material to be touching the exterior of a masonry chimney at *any* location. The Dutch require an additional 25 millimeters (about 1 inch) of insulation to be applied to the exterior of standard types of insulated prefabricated chimneys, and then require 100 millimeters (about 4 inches) of clearance to combustibles instead of our 2 inches.

2. *Is the standard too strict?* To me this question means, "Does the standard require extra costs to be incurred by the consumer with no commensurate safety benefits?" The tougher the standard, the more will be the cost of the chimneys. The standard is overly strict if the allowed temperature on nearby wood is unnecessarily low, or if the flue gas temperature/time inputs are in significant excess of those likely to be encountered in the field with significant frequency. These are difficult issues to resolve. Little laboratory data or statistical field data are available on "low" temperature spontaneous ignition of wood (See Appendix 1).

One can make a case that the temperature inputs in the standard are not strict enough. Flue gas input temperatures in the field certainly can exceed 1000° F., particularly from solid fuel heaters. However, usually such high temperatures are of short duration. It is not known with what probability temperatures may exceed 1000° F. for hours at a time. During chimney fires temperatures higher than 1700° F. sometimes occur.

But the fact that higher flue gas temperatures/durations *can* occur in the field does not imply the existing standard is inadequate. The statistical frequency of these more severe conditions needs to be known. Then the actual accident implications of these conditions must be assessed. Only then can the cost effectiveness of raising the standard be determined. And in a broader perspective, alternative solutions such as *preventing* high flue gas temperatures should also be considered.

3. *Should the standard be applied to masonry chimneys?* Surely any standard for maximum allowable temperature rises on nearby combustibles should be independent of chimney type. So also should the input flue gas temperatures be the same since the chimneys are intended to be used with the same types of appliances. The one area where test procedures might legitimately differ or perhaps be changed for both types of chimneys, is the durations for each flue-gas temperature input. These time durations are especially important for masonry chimneys; their large mass means it takes a longer time for their exterior temperatures to respond to interior temperatures. It takes only a few hours for a metal chimney to warm up to steady-state temperatures, but roughly 24 hours is required for masonry chimneys. Thus the requirement of maintaining 1000° F. flue-gas temperatures *until steady-state* would be unfair to masonry chimneys if 24 hours of continuous 1000° F. input is very rare in the real world. The inherently probabilistic nature of safety is apparent. In any case I doubt that the standard would have specified steady-state conditions if testing masonry chimneys had been intended, merely because of the inconvenience and hence cost of running such long duration tests. But again the hard statistical facts necessary to determine whether the present standard should apply to masonry chimneys are not available.

Another part of the safety standard for factory-built chimneys requires that the chimney itself not be significantly damaged by the test. Here again masonry chimneys would seem to fail. The part of the test that causes the most damage to masonry chimneys is the thermal shock test. This involves introducing approximately 1700° F. flue-gases into a room temperature chimney for 10 minutes. This procedure is repeated two more times with 4 hours between repeats.

All standard masonry chimney types—brick, concrete block, double-thick brick, unlined and lined—crack during such thermal shock tests. Even steady 1000° F. flue-gases crack masonry chimneys, and some types crack at as low as 500° F. Fireclay liners usually crack at temperatures significantly below 1000° F.

Is this cracking dangerous? The answer is not clear. The cracks are often only hairline cracks after the chimney cools, but can be ⅛-inch wide when hot. However under normal conditions (*not* chimney fires) the draft inside the chimney pulls air in through any cracks—hot gases do not normally leak out. Rarely does any fragment of a liner or the masonry fall out. The air tightness of chimneys is decreased by a few times by the thermal shock tests. But does all this make the chimneys unsafe? Again the only way to answer the question is with field evidence, and the needed investigations have not been done. It would appear that this cracking is either not too serious or the required temperatures to crack masonry chimneys are not common, otherwise I would think more fires attributed to normal use of masonry chimneys would be occurring. In any case, keeping masonry chimneys clean so as to avoid chimney fires should avoid much of the cracking and most house fires that could be caused by a cracked chimney.

If the safety standard for factory-built chimneys *is* reasonably applicable to masonry chimneys, and if masonry chimneys fail the test, what is to be done? Code writers ought to look into the issue and consider modifications.

Consumers ought to be cautious, particularly when installing a central wood heater. A thermometer to monitor flue gas temperatures would be a useful accessory for consumers. A barometric draft regulator will help keep temperatures down, and constitutes little additional hazard itself *if* the chimney is kept clean. A check should be made that no wood or other combustible is in direct contact with the chimney. A smoke or heat detector might be installed in the attic if applicable. Any exposed insulation in contact with the chimney should be removed or pulled back. And the chimney should be kept clean enough to avoid chimney fires.

Factory-Built Metal Chimneys

Recently controversy has flared over the safety of factory-built chimneys during chimney fires. A number of house fires have been linked to factory-built chimneys either suffering damage during chimney fires and/or reaching dangerously hot temperatures on their exteriors. Insulated-type factory-built chimneys seem to be more susceptible than do the other types.

During chimney fires flue gas temperatures can exceed the test conditions specified in chimney safety standards. Temperatures can peak well above 2000° F. and may remain above 1700° F. for longer than ten minutes. Apparently some fires have resulted. Unfortunately statistically meaningful data are not yet available that can quantify the seriousness of the situation compared to other wood-heating hazards, including the use of other types of chimneys.

If the hazard is significant, what can be done? Two non-exclusive approaches are to prevent chimney fires from occurring, and to make chimneys safer against chimney fires.

Education of people who heat with wood concerning the importance of maintaining a clean chimney is a voluntary-type preventive approach. Mandatory inspection, and cleaning if necessary, could be legislated. The inspections might have to occur a few times a year for this method to have a large impact. Or the insurance industry could be the moving force by stipulating that chimney-fire-related losses are covered only if there has been some kind of periodic inspection.

Making chimneys safer against chimney fires would involve changing the factory-built-chimney test standard to encompass a larger fraction of field chimney fire conditions—the test temperatures and/or durations could be raised. This would result in safer and more expensive chimneys.

It is not obvious that this option of changing the standards for chimneys is superior to a preventive approach. The extra cost of safer chimneys might not be worth it. This approach does nothing to improve the safety of all the existing chimneys. Also, a safe chimney does not eliminate all chimney fire hazards. Overheated stovepipe and mechanically unsound stovepipe can cause chimney-fire-related house fires, as can the burning debris that spews out the top of chimneys during chimney fires. A preventive program would decrease these causes of fires as well.

Since chimneys that would satisfy a tougher standard are not now readily available in appropriate sizes, a consequence of raising the standards now might be wider use of masonry chimneys. Before this is done, its consequences should carefully be investigated. As discussed previously, masonry chimneys may also have problems. After only about three hours of use of a masonry chimney by 1000° F. flue gases, temperatures of wood around it exceed safe limits (about 200° F.). However, during a chimney fire, masonry has some advantage due to its mass. The outside of a masonry chimney does not respond quickly to changes in flue gas temperatures because of heat storage in the mass. Without more field evidence and laboratory testing it is not clear which type of chimney, on balance, is safer.

It is ironic that the type of prefabricated chimney that appears to be most susceptible to chimney fire damage, the insulated type, is also the one that seems to have the least tendency to accumulate creosote. If all these reputed facts are true, the explanation seems clear. Any means such as insulation that keeps the flue gases warm during normal operation will also necessarily prevent the inner metal liner from dissipating heat during chimney fires, with the possible consequence of liner distortion. This is one more example of a safety tradeoff. Use of insulated chimneys may result in fewer and/or less intense chimney fires due to less creosote accumulation, but if a big chimney fire occurs, damage to the chimney and hence the house may be more likely. As is true in so many other cases, *keeping the flue clean* solves the problem.

Conclusions

So what kind of chimney is best? Masonry chimneys as usually built may not have enough clearance to combustibles during normal high-temperature continuous operation. But short-duration temperature surges are absorbed by the mass with little effect on exterior temperatures. However, masonry chimneys can also crack during chimney fires and their creosote accumulation rate is relatively high.

All listed factory-built chimneys seem to handle normal use relatively safely. But intense chimney fires can apparently cause dangerously high temperatures on nearby combustibles. Insulated factory-built chimneys do not seem to accumulate as much creosote as other types; thus chimney fires may be less frequent and less severe. But intense chimney fires can damage this type of chimney. Factory-built air-insulated or air-cooled (thermosyphon) chimneys appear not to be as susceptible to damage during chimney fires. But they also seem to accumulate more creosote than the insulated chimneys; this seems to be particularly true of the air-cooled type.

Obviously more research is needed. Meantime it is clear that all code-approved chimneys—metal and masonry—are reasonably safe, that no standard residential chimney is absolutely safe, particularly during a chimney fire, and that all chimney fires are avoidable through proper operation and maintenance.

147

APPENDIX 4

The Philosophy of Safety Codes and Standards

Building codes and product safety standards inevitably involve compromises between safety and usefulness, safety and cost, safety and enforceability, and even between safety in one area and safety in another area.

For instance, it would be desirable for the use of wood heaters to result in no creosote accumulation in chimneys. One way to achieve this is to have the smoke in the chimney reach very high temperatures. But this means a large heat loss up the chimney and less heat output of the device into the house—an inefficient heater. Another way to minimize creosote accumulation would be to prohibit airtight stoves. Stoves that always admit some air will burn the smoke more completely and result in less creosote. However, non-airtight stoves usually cannot operate at low power outputs, in some cases use more wood to produce the same amount of total heat output, and always make shutting down the stove less effective as a means of suppressing a chimney fire. Thus non-airtight stoves tend to be less convenient, less efficient, and less safe in certain ways.

As another example, should barometric draft regulators be required as part of all solid-fuel central heater and stove installations? The argument for is that the device effectively limits combustion rates in the fire and flue gas temperatures in the chimney—both desirable objects for safety. On the other hand, the devices may also result in higher rates of creosote accumulation, and more intense chimney fires when they occur. Thus it is unclear whether barometric draft regulators should be required or banned for use with all or some solid-fuel heating systems.

Safety standards for stoves now include limitations on flue gas temperatures. In order to keep flue gas temperatures down, some manufacturers have had to limit the maximum amount of combustion air that can enter the stove by putting a permanent stop on the air inlet damper. The net effect may be counterproductive for safety. So little air can enter the stove that either creosote accumulation increases, or frustrated users operate the units with the doors slightly open, especially when burning green wood or at high altitudes. The hazard of hot coals and sparks falling out of the stove, and of a runaway fire due to too much air, are increased. Thus it is not clear whether there will be fewer or more house fires due to this feature of stove safety standards.

Even if one knew with certainty all the safety aspects of wood heating systems, installation codes intended to protect against every conceivable accident would be undesirable. For instance, standard domestic chimneys, both masonry and factory-built metal, are damaged by intense chimney fires. If codes required chimneys designed to withstand frequent big chimney fires, the cost of such chimneys might be a few thousand dollars, which most people could consider impractical. In this case a preventive approach through education or required inspections and cleanings would be more reasonable.

Codes must be *reasonable*; otherwise they can be counterproductive. If a large fraction of the public perceives a code to be unreasonable, many people will proceed with their wood heater installations on their own without the assistance or recommendations of the local building inspector. The result of an overly restrictive code can thus be less-safe installations. Workable codes require a consensus that they are reasonable.

Ideally safety standards should have a quantitative basis and emphasis. One bird does not a flock make.

148

That Joe Doe's house burned down because a hot round draft control knob fell off its bolt and rolled off the non-level floor protector and ignited the carpet does not mean round knobs on stoves with hot doors should be banned for installation on non-level floors near carpets. Codes would be unenforceably restrictive and complicated if they tried to address every conceivable kind of accident. And the cost of providing extra margins of safety in every installation could easily far exceed the value of property saved by having prevented these rare accidents. Even saving human lives is not worth endless expenditures. Ideally careful accident studies should be used to determine the relative probability of all types of wood-heater related accidents; then codes and safety professionals should emphasize those safety features that are known to be most critical. A quantitative accident probability level should be used to decide both what kinds of safety features to incorporate and to what degree. Safety standards should not be encumbered with minor issues without direct and important safety impacts.

Installation codes need to be simple enough to be easily enforceable. For example, for the same degree of safety, not all types and sizes of stoves need to be installed with the same clearances to combustible walls. Smaller stoves can be closer and stoves with cooler sides can be closer. But to try to write into a building code all the different types of stoves would essentially be impossible, particularly with new stove designs constantly being created. Thus most codes have only one or two clearances that are used to apply to *all* types and sizes of stoves, and they must be adequate for the biggest and hottest stoves. Thus the clearances are excessive for most stoves. This is the price paid for simplicity in the codes, a desirable feature.

But it frustrates well-informed homeowners when a building inspector does not allow an installation because it does not meet the letter of the code even though it is clear the installation is safe. This situation can be equally frustrating for the building inspector. No one is at fault. Regulation cannot be a wholly logical, smooth, and foolproof process.

Codes can stifle innovation, also in the cause of simplicity and hence enforceability. For instance, most codes specify that stoves may be installed with *reduced* clearances to combustible walls if the wall is appropriately protected. The codes list a few specific wall protection systems that allow reduced clearances. The list is very prescriptive—exact materials and installations are specified. This makes the code easy to enforce. But in reality, there are many other suitable materials and systems for protecting combustible walls. If the code were written more generally, to describe the *function* that must be served by the protector, all functionally equivalent systems could be used. This would allow more innovation and esthetic variety. However, who would determine whether alternative systems were in fact equivalent? The local

building inspector rarely has the time or training necessary to make such judgments. Thus, codes often require particular protectors and do not allow any others despite the fact they would indeed work.

Codes and standards implicitly presume a certain level of competence on the part of the operator of the equipment. The chosen level is, of course, somewhat arbitrary. Standards for chimneys presume that enough operators will fail to keep their chimneys clean that chimneys ought to be safe against a chimney fire. Few people would argue with this presumption. At the opposite extreme is the presumption that the operator and other occupants of the home do not know that radiant stoves are hot. Some product standards require prominent labels on stoves saying

---

C A U T I O N

Hot While In Operation
Do Not Touch, Keep Children, Clothing and
Furniture Away

Contact May Cause Skin Burns
See Nameplate and Instructions

Such labeling it not totally without merit—there are people so unfamiliar with wood and coal heaters that they might not know the heaters were hot—but the presumed competence of the operator is clearly low. He is presumed not only not to know that operating stoves are hot, but also not to notice the intense heat radiation from the stove as a clue that the appliance is hot. He is also presumed to notice and understand the label before getting burned. These presumptions may be inconsistent.

For code and standard writers, what operator competence level to presume is a difficult choice. Against severe accidents such as explosions of boilers and water heaters, the system probably should be idiot-proof—no valves should separate the boiler from its pressure relief valve no matter how carefully the instructions might spell out not to close such valves when the boiler is hot. But presuming minimal competence in all situations is not the best solution. Again, costs and benefits should be in proper balance. One way to make stoves perfectly safe is to weld the doors shut so they cannot be used, but here the sacrifice in performance is more than most people would tolerate.

If the worst imaginable succession of incompetent actions is to be rendered harmless by a product standard, the resulting extra cost of the product to all consumers may be way out of proportion to the losses in the rare accidents that would otherwise occur. If installation and operation instructions supplied by manufacturers were to be required by product standards to cover all imaginable contingencies, the instructions might become so long and complicated that they would not be read very carefully if at all, and this would be counterproductive.

Building codes and product safety standards are difficult to write. They require judgments in fuzzy areas and they require compromises. They are intended to result in reasonable safety and practicality. They are never perfect and they are always undergoing revision. But they generally represent the best considered opinions of the experts, with a delay of a few years, since this is the time typically required to change standards and codes.

APPENDIX 5

Safety-Related Products

In this appendix are listed some products and materials that may be useful for the safety of wood and coal heating systems.

Chimney fire extinguisher (looks like a roadside emergency flare)

Chimfex, made by Standard
Railway Fuse Corp.,
Signal Flare Division
P.O. Box 178
Boonton, NJ 07005

Flue gas thermometers

Abbean Cal, Inc.
123-02A Gray Ave.
Santa Barbara, CA 93101

Valley Products and Design, Inc.
P.O. Box 396
Milford, PA 18337

Surface thermometers (usable on stoves and stovepipe)

Condar Co.
Box 264
Garrettsville, OH 44231

Valley Products and Design, Inc.
P.O. Box 396
Milford, PA 18337

Chimney cleaning brushes and related equipment (also available from many woodstove and hardware stores)

Ace Wire Brush Co., Inc.
(manufacturer)
30 Henry St.
Brooklyn, NY 11201

Industrial Construction Co., Inc.
(manufacturer)
765 Conger St.
Eugene, OR 97402

Schaefer Brush Manufacturing
Co. (manufacturer)
117 W. Walker St.
Milwaukee, WI 53204

Worcester Brush Co.
(manufacturer)
38 Austin St.
Worcester, MA 01601

Anchor Tools and Woodstoves
618 N.W. Davis
Portland, OR 97209

Black Magic Chimney Sweeps
Box 977
Stowe, VT 05672

Dan Patch Chimney Sweeps,
Inc.
2830 Cedar Ave., South
Minneapolis, MN 55407

Hearth Enterprises, Inc.
508 Shorter Ave.
Rome, GA 30161

Ken Hinkley, Master Sweep
Box 180
Williamsburg, MA 01096

Kristia Associates
P.O. Box 1118
Portland, Maine 04104

Stainless steel stovepipe

The Dalsin Manufacturing Co.
8824 Wentworth Ave., South
Minneapolis, MN 55420

Warm Up Stove Co.
P.O. Box 51
Helmsburg, IN 47435

Bow and Arrow Stove Co.
11 Hurley St.
Cambridge, MA 02141

Condensation Engineering Corp.
3511 West Potomac Ave.
Chicago, IL 60651

Enamel-coated steel pipe

Bow and Arrow Stove Co.
11 Hurley St.
Cambridge, MA 02141

Chimney fire alarm

Valley Products and Design, Inc.
P.O. Box 396
Milford, PA 18337

Wall protector spacing and mounting kit

Veneered Metals, Inc.
Box 327
Edison, NJ 08817

Heavy-gauge stovepipe

Dura-Vent Corp.
2525 El Camino
Redwood City, CA 94064

Sunrise Energy Products
996 Mackinaw Highway
Pellston, MI 49769

Thompson and Anderson, Inc.
446 Stroudwater St.
Westbrook, ME 04092

Furnace cement

Industrial Gasket and Shim Co.,
Inc.
Meadow Lands, PA 15347

Heat resistant gloves

Apex Glove Co., Inc.
3820 West Wisconsin Ave.
Milwaukee, WI 53208

Duct insulation and insulated duct

Many building supply stores sell
these materials. Insulation rated
for use only up to 250° F. should
not be used within 6 feet of the
plenum of a hand-fired solid-fuel
furnace.

Babcock and Wilcox
Refractories Division
P.O. Box 923
Augusta, GA 30903

Johns-Manville
Ken-Caryl Ranch
Denver, CO 80217

*Hot coal catcher and ash pan for
many Scandinavian-style stoves*

Wood Structures
Box 27
Brinklow, MD 20727

*High-temperature insulation (such
as ceramic wool in bulk, bat and
rope forms)*

Babcock and Wilcox Co. (makers
of Kaowool)
P.O. Box 923
Augusta, GA 30903

The Carborundum Co. (makers
of Fiberfrax)
P.O. Box 337
Niagara Falls, NY 14302

Refractory Products Co.
Box 309
Carpentersville, IL 60110

Sauder Industries, Inc.
220 Weaver St.
Emporia, KA 66801

APPENDIX 6

Miscellaneous Technical Data

Table A6-1. Relation of Gauge Number to Thickness

Gauge No.	Thickness (Inches)	Gauge No.	Thickness (Inches)	Gauge No.	Thickness (Inches)	Gauge No.	Thickness (Inches)
3	0.239	12	0.105	21	0.033	30	0.0120
4	0.224	13	0.090	22	0.030	31	0.0105
5	0.209	14	0.075	23	0.027	32	0.0097
6	0.194	15	0.067	24	0.024	33	0.0090
7	0.179	16	0.060	25	0.021	34	0.0082
8	0.164	17	0.054	26	0.018	35	0.0075
9	0.150	18	0.048	27	0.016	36	0.0067
10	0.135	19	0.042	28	0.015	37	0.0064
11	0.120	20	0.036	29	0.0135	38	0.0060

Adapted from T. Baumeister, ed., *Mark's Standard Handbook for Mechanical Engineers*, 7th ed. (New York: McGraw-Hill, 1967), p. 6–48.

There are at least six gauge scales. This table gives thickness according to the Manufacturers' Standard Gauge (MSG), which is commonly used for uncoated steel sheets. Other gauge scales differ from MSG by no more than about 10 percent, except for AWG (or B & S), which is used principally for nonferrous sheets, rod and wire (e.g., copper wire).

Table A6-2. Relation of Color to Temperature of Iron or Steel

	°F.	°C.		°F.	°C.
Dark blood red, black red	990°F.	530°C.	Orange	1650°F.	900°C.
Dark red, blood red, low red	1050	565	Light orange	1725	940
Dark cherry red	1175	635	Yellow	1825	995
Medium cherry red	1250	675	Light yellow	1975	1080
Cherry, full red	1375	745	White	2200	1205
Light cherry, light red	1550	845			

Adapted from T. Baumeister, ed., *Marks' Standard Handbook for Mechanical Engineers*, 7th ed. (New York: McGraw-Hill, 1967), p. 4–7.

Table A6-3. NFPA Protector Extents for Appliances

Permitted Clearance with No Protector[1] (Inches)	Reduced Clearance[2] (Inches)	Distance from Wall to Front Surface of Protector (Inches)				
		0	1	2	4	6
36	18	31.2	29.4	27.7	24.2	20.8
	12	33.9	31.1	28.4	24.3	20.8
18	9	15.6	13.5	12.1	9.1	6.7
	6	17.0	14.2	12.1	9.1	—[3]
	3	17.7	14.2	12.1	—	—
12	9	7.9	7.1	6.2	4.4	2.7
	6	10.4	8.7	7.0	4.5	—[3]
	4	11.3	8.7	7.0	—[3]	—
	2	11.8	8.7	—[3]	--	—

1. See Table 2-4. The permitted clearance with no protection is determined by the type of appliance and, in some cases, its orientation.

2. See Table 2-5. Given the appliance, the permitted reduced clearance is determined by the type of protector.

3. Although it is not prohibited by NFPA, the author does not recommend direct contact between protector and appliances. These entries have been left blank.

Above are NFPA's and most building codes' minimum wall protector extents for appliances.

The numbers in the table are not the full widths of the required protectors but represent how much wider on each side and how much higher the protector must be than the appliance itself. Thus the protector must extend the indicated number of inches above the top of the appliance and must be wider than the appliance by twice the indicated number. This table also applies to ceiling protectors.

These protector extents are computed from information in NFPA 89-M, "Heat-Producing Appliance Clearances 1976," under the assumptions that the appliance is rectangular and is oriented parallel or perpendicular, but not out of line with the protected wall. For most other shapes these computed protector extents are slightly more than NFPA requires. These clearances do not apply to askew installations such as may occur when a fireplace stove is installed in a corner.

Technical note: Measuring the no-protection clearance from the corner of the stove does not guarantee compliance with NFPA 89-M. For protectors that extend out from the wall, the protector width must be adequate to assure that all of the stove's side is also sufficiently far from unprotected wall surface. Proper use of the table does assure compliance with NFPA 89-M.

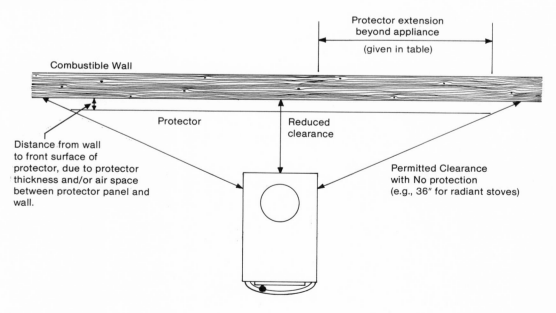

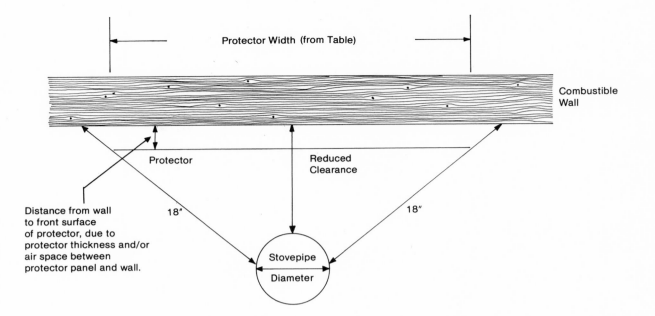

Table A6-4. NFPA Stovepipe Protector Widths

5-INCH (DIAMETER) STOVEPIPE

Reduced Clearance[1] (Inches)	Distance from Wall to Front Surface of Protector (Inches)				
	0	1	2	4	6
12	29.0	27.0	25.1	21.3	17.9
9	33.9	31.0	28.3	23.1	18.3
6	37.3	33.1	29.3	22.3	—[2]
3	39.5	33.2	27.8	—	—

6-INCH STOVEPIPE

Reduced Clearance[1] (Inches)	Distance from Wall to Front Surface of Protector (Inches)				
	0	1	2	4	6
12	29.4	27.5	25.6	22.0	18.6
9	34.4	31.7	29.0	24.0	19.4
6	37.9	33.9	30.3	23.7	—[2]
3	40.2	34.3	29.3	—	—

---

1. See Table 2-5. The allowed reduced clearance depends on the type of protector used and on whether a wall or ceiling is being protected.
2. Although not prohibited by NFPA, the author does not recommend direct contact between stovepipe and protector. Thus this entry has been left blank.

NFPA's (and most building codes') minimum widths of protector panels for reduced clearances from stovepipes to parallel walls or ceilings. These widths have been computed from information in NFPA 89-M, "Heat-Producing Appliance Clearances 1976," and give protection to all wall surface within 18 inches from the stovepipe. (Technical note: measuring the 18 inches from the *nearest* part of the stovepipe (where the pipe's surface is perpendicular to the 18-inch line to the wall) does not necessarily assure other parts around the circumference

7-INCH STOVEPIPE

Reduced Clearance[1] (Inches)	Distance from Wall to Front Surface of Protector (Inches)				
	0	1	2	4	6
12	29.8	27.9	26.1	22.6	19.3
9	35.0	32.3	29.6	24.8	20.5
6	38.5	34.7	31.2	24.9	—[2]
3	41.0	35.4	30.7	—	—

8-INCH STOVEPIPE

Reduced Clearance[1] (Inches)	Distance from Wall to Front Surface of Protector (Inches)				
	0	1	2	4	6
12	30.2	28.3	26.5	23.1	19.9
9	35.5	32.8	30.3	25.6	21.4
6	39.2	35.5	32.1	26.0	—[2]
3	41.7	36.4	32.0	—	—

10-INCH STOVEPIPE

Reduced Clearance[1] (Inches)	Distance from Wall to Front Surface of Protector (Inches)				
	0	1	2	4	6
12	31.0	29.2	27.5	24.2	21.2
9	36.5	34.0	31.6	27.2	23.2
6	40.4	36.9	33.7	28.1	—[2]
3	43.1	38.3	34.2	—	—

---

of the pipe are more than 18 inches away from unprotected wall. This effect is most pronounced for protectors that extend out considerably from the wall; the table incorporates this effect. (Since NFPA protector thicknesses are minimal, NFPA protectors can extend a few inches out from the protected surface.))

Bibliography

Books on Heating with Wood

Adkins, Jan. *The Art and Ingenuity of the Wood Stove.* New York: Everett House, 1978.

Daniels, M.E. *Fireplace and Wood Stoves.* Indianapolis and New York: Bobbs-Merrill, 1977.

Gay, Larry. *The Complete Book of Heating with Wood.* Charlotte, Vt.: Garden Way Publishing, 1974.

Havens, David. *The Woodburners Handbook.* Brunswick, Maine: Harpswell Press, 1973.

Ivins, David. *The Complete Book of Woodburning Stoves.* New York: Drake Publishers, 1978.

Ross, Bob and Carol. *Modern and Classic Woodburning Stoves.* Woodstock, N.Y.: Overlook Press, 1976.

Shelton, Jay. *The Woodburners Encyclopedia.* Waitsfield, Vt.: Vermont Crossroads Press, 1976.

Soderstrom, Neil. *Heating Your Home with Wood.* New York: Popular Science/Harper, 1978.

Twitchell, Mary. *Wood Energy.* Charlotte, Vt.: Garden Way Publishing, 1978.

Vivian, John. *Wood Heat.* Emmaus, Pa.: Rodale Press, 1976.

Wilk, Ole. *Wood Stoves: How to Make and Use Them.* Anchorage, Alaska: Alaska Northwest Publishing Co., 1977.

Periodicals

Home Energy Digest and Wood Burning Quarterly, 8009, 34th Avenue South, Minneapolis, Minnesota 55420.

The National Wood Stove and Fireplace Journal, 10221 Slater, Suite 209, Fountain Valley, California 92708.

Chimney Cleaning

Curtis, Christopher, and Post, Don. *Be Your Own Chimney Sweep.* Charlotte, Vt.: Garden Way Publishing, 1979.

Owner-Built Fireplaces

Eastman, Margaret and Wilbur F., Jr. *Planning and Building Your Fireplace.* Charlotte, Vt.: Garden Way Publishing, 1976.

Kern, Ken, and Magers, Steve. *Fireplaces.* Oakhurst, Calif.: Owner-Builder Publications, 1978.

Water Heating with Wood

Sussman, Art, and Frazier, Richard. *Handmade Hot Water Systems.* Point Arena, Calif.: Garcia River Press, 1978.

Safety Pamphlets and Booklets

Burning Wood. Ithaca, N.Y.: Northeast Regional Agricultural Engineering Service, 1977.

Heating with Wood—Safely. Canada: Central Mortgage and Housing Corporation, 1978.

Using Coal and Wood Stoves Safely. NFPA No. HS-10. Boston: National Fire Protection Association, 1978.

Wood for Home Heating. (A series of pamphlets.) Madison, Wis.: Wisconsin Energy Extension Service, 1978.

Wood Fuel Heating Tips. Madison, Wis.: American Family Mutual Insurance Co., 1978.

Safety Standards and Related Materials

Underwriters Laboratories, Inc., 333 Pfingsten Road, Northbrook, Ill. 60062:

UL 737	Fireplace Stoves
UL 103	Chimneys: Factory-Built, Residential Type and Building Heating Appliance
UL 127	Factory-Built Fireplaces
UL 1482	Solid-Fuel-Type Room Heaters

| UL 462 | Proposed Standard for Heat Reclaimers for Gas-, Oil- or Solid Fuel-Fired Appliances |

National Fire Protection Association, 470 Atlantic Avenue, Boston, Mass. 02210

NFPA 89M	Heat-Producing Appliance Clearances, 1976.
NFPA 97M	Glossary of Terms Relating to Heat Producing Appliances, 1972.
NFPA 211	Chimney, Fireplaces and Vents, 1977. (Revisions, especially concerning stove installations, are currently under review by NFPA.)
NFPA 90B	Warm-Air Heating and Air Conditioning Systems, 1976.

Fire Protection Handbook, 14th edition

Model Building Codes

Canadian Heating, Ventilating and Air-Conditioning Code 1977. Ottawa, Canada: National Research Council of Canada, 1977.

The BOCA Basic Mechanical Code/1978. Chicago: Building Officials and Code Administrators International, Inc., 1978.

The National Building Code. New York: American Insurance Association, 1976.

The Standard Building Code. Birmingham, Ala.: Southern Building Code Congress International, 1976.

Uniform Mechanical Code. Whittier, Calif.: International Association of Plumbing and Mechanical Officials, and International Conference of Building Officials (ICBO), 1976.

Studies of Fires Related to Wood Heating

Lawson, D.I.; Fox, L.L.; and Webster, C.T. *The Heating of Panels by Flue Pipes.* Fire Research Special Report No. 1. England: Department of Scientific and Industrial Research, 1952.

Peacock, Richard D. *A Review of Fire Incidents, Model Building Codes, and Standards Related to Wood-Burning Appliances.* Washington, D.C.: Center for Fire Research, National Engineering Laboratory, National Bureau of Standards, 1978.

Shelton, Jay W. "Analysis of Fire Reports on File in the Massachusetts State Fire Marshall's Office Relating to Wood and Coal Heating Equipment." NBS-GCR-78-149. Washington, D.C.: Center for Fire Research, National Engineering Laboratory, National Bureau of Standards, 1978.

Research on the Safety of Solid-Fuel Heater Installations

Clearances and Insulation of Heating Appliances. Bulletin of Research No. 27. Northbrook, Ill.: Underwriters Laboratories, Inc., 1943.

Voigt, G.Q. "Fire Hazard of Domestic Heating Installations." Research Paper RP 596. *Bureau of Standards Journal of Research*, September 1933, pp. 353–72.

Safe Temperature Limits of Wood

Browne, F.L. *Theories of the Combustion of Wood and Its Control.* Report No. 2136. Madison, Wis.: Forest Products Laboratory, 1963.

Ignition and Charring Temperatures of Wood. Report No. 1464. Madison, Wis.: Forest Products Laboratory.. Also printed in *Wood Products* 50 (1945) 21–22.

Jackman, P.E., and Saunders, R.G. *The Fire Performance of Timber—A Literative Survey.* Section 3, Theories of Ignition. Timber Research and Development Association, May 1972.

Lawson, D.I. "The Ignition of Wood by Radiation." *British Journal of Applied Physics* 3 (1952), pp. 288–92.

McGuire, J.H. "Limited Safe Surface Temperature of Combustible Materials." *Fire Technology*, August 1969, pp. 237–41.

Mitchell, N.D. "New Light on Self-Ignition." *NFPA Quarterly*, October 1951.

"Survey of Available Information on Ignition of Wood Exposed to Moderately Elevated Temperatures." Part II of *Performance of Type B Vents for Gas-Fired Appliances*, Research Bulletin No. 51. Northbrook, Ill. Underwriters Laboratories, Inc., 1959.

Masonry Chimney Safety

"Chimneys, Tests on Masonry Residential Type." Special Interest Bulletin No. 14. National Board of Fire Underwriters, May 1960.

Fields, E.F. "Fire Hazard Test Made on a Standard 7 Brick Lined Chimney Under Conditions of UL Subject 103." Report No. 42, Project No. 107–A. Belmont, Calif.: William Wallace Company, October 1955.

Mitchell, N.D. "Fire Hazard Tests With Masonry Chimneys." *NFPA Quarterly*, October 1949.

Shoub, H. "Survey of Literature on the Safety of Residential Chimneys and Fireplaces." NBS Miscellaneous Publication 252. Washington, D.C.: National Bureau of Standards, December 1963.

Tamura, G.T., and Wilson, A.G. "Fire Hazard Tests on Small Masonry Chimneys." Internal Report No. 202, Division of Building Research. Ottawa, Canada: National Research Council, December 1960.

Thulman, R.K. "Performance of Masonry Chimneys for Houses." Housing Research Paper No. 13, Division of Housing Research. Washington, D.C.: Housing and Home Finance Agency, November, 1952.

Wachmann, C. "An Annotated Bibliography on Residential Chimneys Serving Solid- or Liquid-Fuel Fired Heating Appliances." Division of Building Research. Ottawa, Canada: National Research Council, April 1957.

Factory-Built Chimney Safety

Flink, Carl R. "Factory-Built Chimneys." *Proceedings of Wood Heating Seminar 4*, Camden, Maine: Wood Energy Institute, 1979. pp. 339–78.

Performance of Wood (and Coal) Heating Equipment

American Society of Heating, Air-Conditioning and Refrigerating Engineers. *ASHRAE Guide and Data Book, 1975 Equipment Volume,* Chapter 26. New York.

Bastings, L., and Benseman, R.F. "Calorimetry of Solid-Fuel Domestic Stoves." *Journal of the Institute of Heating and Ventilating Engineers,* July 1953.

Bastings, L. and Benseman, R.F. "Solid-Fuel-Burning Domestic Stoves." *New Zealand Engineering,* 15 May 1952.

Bull, Marcus. "Experiments to Determine the Comparative Quantities of Heat Evolved in the Combustion of the Principal Varieties of Wood and Coal Used in the United States for Fuel, and, Also, to Determine the Comparative Quantities of Heat Lost by the Ordinary Apparatus Made for Their Combustion." *Transactions of the American Philosophical Society III.* New Series, 1930.

Eaton, F.J. "Rating of Domestic Solid-Fuel Appliances." *Gas Journal,* June 1951, pp. 833–84.

Fishenden, M.W. "The Domestic Grate." Technical Paper No. 13, Fuel Research Board. London: Department of Scientific and Industrial Research, 1925.

Fishenden, M.W. "The Efficiency of Low-Temperature Coke in Domestic Appliances." Technical Paper No. 3, Fuel Research Board. London: Department of Scientific and Industrial Research, 1921.

Fishenden, M.W. "The Coal Fire." Special Report No. 3, Fuel Research Board. London: Department of Scientific and Industrial Research, 1920.

Fox, L.L. "Efficiency of Domestic Space-Heating Appliances Using Solid Fuel." *Journal of the Institute of Fuel* 25 (1952), pp. 267–76. England.

"Improving the Efficiency, Safety and Utility of Woodburning Units." Quarterly Reports. Auburn, Ala.: Department of Mechanical Engineering, Auburn University.

Kollock, T. "Efficiency of Wood in Stove and Open Fireplaces." *Forest, Fish and Game* 3 (1911), pp. 95–97.

Konzo, S. and Harris, W.S. "Fuel Savings Resulting From Closing of Rooms and From Use of a Fireplace." Bulletin Series No. 348. University of Illinois Engineering Experiment Station, 16 November 1943.

Landry, B.A. and Sherman, R.A. "The Development of a Design of Smokeless Stove for Bituminous Coal." *Transactions of the ASME* 72 (1950), p. 9.

Rowley, F.B. and Allen, J.R. "Tests to Determine the Efficiency of Coal Stoves." *Transactions of the ASHVE* 26 (1920), p. 115–22.

Rowse, R.H. and Moss, W.C. "Domestic Boilers and Stoves Using Solid Fuel." *Journal of the Intitution of Heating and Ventilating Engineers* 18 (1950), England, pp. 32–78.

Seeley, L.E. and Keator, F.W. "Wood-Burning Space Heaters." *Mechanical Engineering* 62 (1940), p. 864.

Shelton, J.W. "An Analysis of Woodstove Performance." *Blair and Ketchum's Country Journal,* October 1976, pp. 47–50.

Shelton, J.W. "Measured Performance of Fireplaces and Fireplace Accessories." Williamstown, Mass: published by Jay Shelton, 1979. An abridged version of this paper was published as "What Fireplaces and Fireplace Accessories Are Really

Worth." *Home Energy Digest and Wood Burning Quarterly*, Winter 1978.

Shelton, J.W. "Steadiness and Control in Wood Heating Systems." *Wood Burning Quarterly*, Summer 1977.

Shelton, J.W., Black, Chaffee and Schwartz. "Wood Stove Testing Methods and Some Preliminary Results." *ASHRAE Trans.* 48 (1978), pp. 388–404.

Stone, R.L. "Fireplace Operation Depends on Good Chimney Design." *ASHRAE Journal*, February 1969.

Note: All publications by the author (J. W. Shelton) are available directly from him. For a current list with prices, send a stamped self-addressed envelope to Dr. Jay W. Shelton, Box 5235 Coronado Station, Santa Fe, New Mexico 87502.

Air Pollution

Boubel, Richard W. "Particulate Emissions from Sawmill Waste Burners." Bulletin No. 42. Corvallis, Oreg.: Engineering Experiment Station, Oregon State University, 1968.

Butcher, Samuel S. "The Air Pollution Potential of Small Wood Stoves." Paper delivered at Wood Heating Seminar 3, April 1978, Madison, Wis. Sponsored by Wood Energy Institute, Camden, Maine.

Kircher, D. "Source Testing for Fireplaces, Stoves, and Restaurant Grills in Vail, Colorado." Kansas City: Pedco-Environmental, Inc., 1977.

Prakash, C.B. and Murray, F.E. "Studies on Air Emissions from the Combustion of Wood Waste." *Combustion Science and Technology* 6 (1972), pp. 81–88.

Seeley, L.E. and Keator, F.W. "Wood-Burning Space Heaters." *Mechanical Engineering* 62 (1940), p. 864.

Smith, J.R. and Suggs, J.C. "Smoke Composition." Proceedings of Internation Symposium on Air Quality and Smoke from Urban and Forest Fires at Fort Collins, Col. National Academy of Sciences, 1976, pp. 296–317.

Smoke Detectors

Fire Detection for Life Safety. Proceedings of a symposium held March 31 and April 1, 1975. National Academy of Sciences, Washington, D.C., 1977.

"Smoke Detectors." *Consumer Reports*, October 1976, pp. 555–59.

Organizations, principally of and for manufacturers, distributors, and dealers:

Canadian Wood Energy Institute
Suite 103A, 55 Isabella St.
Toronto, Ontario M4Y 1M8

Chimney Sweep Guild
c/o Kristia Associates
P.O. Box 1176
Portland, ME 04104

Fireplace Institute
111 East Wacker Dr.
Chicago, IL 60601

Gas Appliance Manufacturers Association
P.O. Box 9245
Arlington, VA 22209

National Solid Fuel Trades Assoc., Inc.
504 Empire Bldg.
Syracuse, NY 13202

Wood Energy Institute
1101 Connecticut Avenue
Suite 700
Washington, D.C. 20036

Safety Organizations

National Fire Protection Association
470 Atlantic Ave.
Boston, MA 02210

Underwriters Laboratories, Inc.
333 Pfingsten Road
Northbrook, IL 60062

Glossary

Airtight stove. A stove in which a large fire can be suffocated by shutting the air inlets, resulting ultimately in a large mass of unburned fuel remaining in the stove.

Appliance (as used in this book). A solid-fuel burning stove, fireplace, furnace, boiler, water heater, or cook stove.

Aquastat. An automatic device for controlling water temperature.

Boiler, hot water. A hot-water central-heating appliance.

Boiler, steam. A steam central heating appliance.

Breaching. Horizontal access into a chimney for a chimney connector. In prefabricated chimneys a T provides a breaching. In masonry chimneys, breachings are holes that should have a sleeve or liner of tile or heavy steel.

Btu (British Thermal Unit). A unit for measuring energy, equal to the amount of energy needed to increase the temperature of 1 pound of water by 1 degree Fahrenheit.

Chimney capacity. The maximum safe venting capability of a chimney, most often expressed in terms of the fuel consumption rate of connected appliances (in Btu per hour), but more fundamentally related to the mass flow (e.g., pounds per minute) of flue gas which will flow up the chimney under given conditions of temperature and barometric pressure.

Chimney connector. The connector between an appliance and its chimney. Stovepipe is commonly used for chimney connectors. Some appliances have no connectors because the chimney connects directly to the appliance (e.g., many prefabricated fireplaces).

Chimney fire. The burning of creosote/soot deposits inside a chimney or stovepipe.

Circulating fireplace. A fireplace with multiple-wall construction around the fire chamber which permits air to circulate between the walls, become heated, and enter the house either directly or via short ducts.

Circulating stove. See stove, circulating.

Collar (or flue collar). The part of a fuel-burning appliance to which the chimney connector or chimney attaches.

Combustible. (as applied to walls, floors and ceilings in the context of wood heater clearances for safety) Constructed of or surfaced with wood, paper, natural- or synthetic-fiber cloth, plastic or other material which will ignite and burn, whether flameproofed or not and whether plastered or unplastered. Combustibility is a relative concept. This definition is adapted from the definition in NFPA booklet "Glossary of Terms Relating to Heat-Producing Appliances." (Boston: National Fire Protection Association), 1972.

Combustion efficiency. The percentage of the total energy content of the fuel consumed that is converted to heat in the fire.

Cook stove. A wood- or coal-burning appliance with a closed fire chamber, which is intended primarily for cooking and includes an oven.

Creosote. Chimney and stovepipe deposits originating as condensed wood smoke (including vapors, tar and soot). Creosote is often initially liquid, but may dry or pyrolyze to a flaky or solid form. Smoldering wood (or coal) is the main source of creosote.

Damper. A valve, usually a movable or rotatable plate, for controlling the flow of air or smoke.

Domestic water. The water piped through a house to basin, sink, shower, and tub outlets. It is presumed potable, as contrasted with "industrial" water.

Draft. The difference in air pressure at the same elevation between the inside and the outside of a chimney, chimney connector, or appliance. The term "draft" is also sometimes used to denote the rate of combustion air flow into a fuel-burning appliance, or the rate of flue gas flow.

Draft regulator, barometric. A device designed to prevent excessive draft in a fuel-burning appliance by admitting air to the venting system. Draft regulators are usually installed in chimney connectors.

Elbow, stovepipe. Stovepipe fittings or sections involving turns or bends. 90-degree elbows are most common. Some types of elbows are adjustable from 90 degrees to 0 degrees (no bend).

Energy efficiency (or overall energy efficiency). The percentage of the total energy content of the fuel consumed that becomes useful heat in the house.

Firebrick. Brick capable of withstanding high temperatures, such as in stoves, furnaces and boilers. Different types of firebrick have different temperature limits.

Fireclay. Clay that will withstand high temperatures without cracking or deforming. NFPA recommends that fireclay chimney flue liners resist corrosion, softening or cracking from flue gases at temperatures up to 1800° F.

Fireplace stove. A free-standing solid-fuel-burning room-heating appliance operated either with its fire chamber open or closed to the room. NFPA and most codes use the term "Room Heater—Fireplace Stove Combination" for a fireplace stove, and use the term "fireplace stove" to designate a unit without doors, that has its firechamber always open to the room.

Fireplace, zero clearance. A factory-built metal fireplace with multi-layer construction providing enough insulation and/or air cooling so that the base, back, and in some cases sides, can safely be placed in direct contact ("zero clearance") with combustible floors and walls.

Flue gases. The gases in an operating venting system, consisting of combustion products plus whatever air is mixed with them. Essentially synonymous with "smoke" and "stack gases."

Fly ash. Ash that goes up the chimney, as opposed to ash that remains in the fuel-burning appliance.

Furnace. 1. (The definition used in this book) A hot-air central-heating appliance. 2. Any central heating appliance. 3. The combustion chamber of any fuel-burning appliance.

Hearth. The floor of the combustion chamber in coal or wood burning appliances.

Hearth extension. The non-combustible co-planar extension of the hearth beyond the opening of a fireplace or fireplace stove. The term is also sometimes used to denote the floor protector under or around any residential solid-fuel-burning appliance.

Heat transfer efficiency. The percentage of heat, which is released in the fire, that gets through the appliance and chimney walls to become useful heat in the house.

Infrared radiation. The invisible and harmless radiation emitted by a hot object. This radiation is converted into heat when it is absorbed.

Insulating brick. Low density (high porosity), low thermal conductivity firebrick intended for use in kilns and furnaces to insulate them, reducing heat losses. Its conductivity and its heat storage capacity are both 1/5 to 1/3 that of hard firebrick.

Liner, chimney. Usually a high-temperature clay ("fireclay") round or rectangular sleeve lining the interior of masonry chimneys. Although not recognized by many building codes, other materials such as stainless steel stovepipe and enameled porcelain-coated steel industrial chimneys can be used as liners.

Liner, stove. A layer of metal or brick placed immediately adjacent to a side or bottom of a stove, intended either to protect the main stove structure from getting too hot, or to insulate the combustion chamber, making it hotter and thus promoting more complete combustion. Liners are usually designed for easy replacement.

Listed. Included in a list published by a recognized testing laboratory or inspection agency, indicating that the equipment meets nationally recognized safety standards.

Moisture content. The percentage of fuel wood weight that is moisture. The moisture content of a piece of wood may be determined by cutting the piece into 1-inch slabs, weighing the slabs, placing them in an oven at about 220° F. for about a day (or until their weight stops changing) and then reweighing them. The dried slabs have zero moisture content (by definition). The lost weight is moisture. Thus the moisture content (in percent) is the weight difference times 100 divided by the original weight.

NFPA. The National Fire Protection Association, 470 Atlantic Avenue, Boston, MA 02210. An independent not-for-profit organization for fire safety.

Primary combustion. The burning of solid wood and some of the combustible gases, which takes place in that portion of the appliance where the wood is. The distinction between primary and secondary combustion is somewhat artificial.

Pyrolysis. The chemical alteration of wood or coal by the action of heat alone, in the absence of oxygen and hence without burning. The products of pyrolysis are gases, tar fog and charcoal or coke.

Radiant stove. See *Stove, radiant.*

Refractory. Any solid ceramic material suitable as a structural or protective material at high temperatures in a corrosive environment.

Seasoned wood. Wood that has lost a significant amount of its original (green) moisture. The term has no both quantitative and universally accepted meaning.

Secondary combustion. The burning of the combustible gases and smoke which are not burned in primary combustion.

Secondary combustion chamber. The place where secondary combustion occurs.

Solid fuel. Fuel in solid form, such as wood, coal, paper and related products.

Soot. Soft black velvety carbon particle deposits inside appliances, chimneys or connectors. Soot originates in oxygen-poor flames.

Stack effect. The effects resulting from the warm air in buildings on a cold day being relatively buoyant, just as are the flue gases in a chimney or stack. Effects include pressure differences between inside and outside the building, airflow into the building in the lower stories and airflow out of the building in the upper portions.

Stove, wood stove or coal stove. A free-standing solid-fuel-burning, room-heating appliance intended to be operated with its door(s) closed, i.e., with a closed fire chamber. NFPA and most codes use the term "solid-fuel room heater" for stoves.

Stove, circulating. A stove with an outer jacket (usually sheet metal) beyond the main structure, with openings at or near the bottom and top so that air can circulate between the stove body and its jacket. For purposes of determining safe clearances, a circulating stove must be fully jacketed on all four sides, including at the access doors and on the top.

Stove, radiant. A stove whose heat output is mostly in the form of radiant energy.

Stove mat (or stoveboard). A prefabricated panel used as a floor or wall protector.

Stovepipe (or smokepipe). Single-walled light gauge (roughly .019 to .024 inches thick) metal pipe generally intended for use as chimney connectors.

Stovepipe damper. A damper to be installed in a stovepipe connector to regulate flow and draft.

Tap water. *See Domestic water.*

Thermostat. An automatic device for regulating the temperature in a building by controlling the heating or cooling source or its distribution.

Thimble. A device to be installed in combustible walls, through which stovepipe passes, intended to help protect the walls from igniting due to stovepipe heat. A thimble by itself is not usually adequate. The simplest thimbles are simply metal or fire clay sleeves or cylinders.

UL. Underwriters Laboratories, Inc., 333 Pfingsten Road, Northbrook, IL 60062. An independent, not-for-profit organization testing for public safety.

Water heater. An appliance intended principally for heating domestic (or tap) water.

Index

Other Garden Way Books You Will Enjoy

The energy-conscious homeowner concerned about alternate fuels will find an up-to-date library essential. Here are some excellent books in these areas.

Be Your Own Chimney Sweep, by Chris Curtis and Donald Post. 112 pages, 6 × 9, quality paperback, $4.95. Two professional sweeps tell how to clean stoves, stovepipes and chimneys efficiently and safely.

Wood Energy: A Practical Guide to Heating With Wood, by Mary Twitchell. Quality paperback, 8½ × 11, 176 pages, $7.95. The definitive wood heat book, with comprehensive catalog section on stoves and furnaces.

Home Energy for the Eighties, by Ralph Wolfe and Peter Clegg. Quality paperback, 8½ × 11, 272 pages, $10.95. How to deal with the energy crisis by turning to solar heat, water power, wind power, and wood. Plus illustrated catalog sections on what's available now in these fields.

At Home in the Sun: An Open-House Tour of Solar Homes in the United States, by Norah Davis and Linda Lindsey. Quality paperback, 8½ × 11, 248 pages, $9.95. What it's really like to live in a solar house, as told by the owners of thirty-one solar homes around the country.

Harnessing Water Power for Home Energy, by Dermot McGuigan. 112 pages, quality paperback, $4.95; hardback $9.95. An authoritative, detailed look at the uses of small-scale water power.

Harnessing Wind Power for Home Energy, by Dermot McGuigan. 144 pages, quality paperback, $4.95; hardback $9.95. A solid, complete analysis of wind power options for homeowners, with details on machines, manufacturers, and whole systems.

Build Your Own Low-Cost Log Home, by Roger E. Hard. 204 pages, 8½ × 11, quality paperback, $6.95; cloth, $10.95. A remarkably complete home construction book.

Designing and Building a Solar House, by Donald Watson. 288 pages, 8½ × 11, quality paperback, $9.95; cloth, $12.95. "A nuts-and-bolts book that brings the sun down to earth," said Alvin Toffler, author of *Future Shock.*

Heating with Wood, by Larry Gay. 128 pages, 6 × 8, quality paperback, $3.95. All the basic information you need for switching to wood heat.

These books are available at your bookstore, or directly from Garden Way Publishing, Box 171X, Charlotte, Vermont 05445. If ordering by mail and your order is under $10, please enclose 75¢ for postage and handling.